Energy Alternatives

Series Editor: Cara Acred

Volume 277

Independence

First published by Independence Educational Publishers

The Studio, High Green

Great Shelford

Cambridge CB22 5EG

England

© Independence 2015

Copyright

Photocopy licence

British Library Cataloguing in Publication Data

Energy alternatives. -- (Issues ; 277)
1. Renewable energy sources.
I. Series II. Acred, Cara editor.
333.7'94-dc23

ISBN-13: 9781861687050

Printed in Great Britain
Zenith Print Group

Contents

Introduction

Energy Alternatives is Volume 277 in the **ISSUES** series. The aim of the series is to offer current, diverse information about important issues in our world, from a UK perspective.

ABOUT ENERGY ALTERNATIVES

With non-renewable energy sources such as coal, gas and oil running out, it is increasingly important for us to examine energy alternatives for the future. This book explores sustainable and renewable energy sources, such as bioenergy, wave and tidal energy, and even poo power! It also considers current statistics about the fuel we use, the debate surrounding fracking and information about energy prices and efficiency.

OUR SOURCES

Titles in the **ISSUES** series are designed to function as educational resource books, providing a balanced overview of a specific subject.

The information in our books is comprised of facts, articles and opinions from many different sources, including:

⇨ Newspaper reports and opinion pieces

⇨ Website factsheets

⇨ Magazine and journal articles

⇨ Statistics and surveys

⇨ Government reports

⇨ Literature from special interest groups.

A NOTE ON CRITICAL EVALUATION

Because the information reprinted here is from a number of different sources, readers should bear in mind the origin of the text and whether the source is likely to have a particular bias when presenting information (or when conducting their research). It is hoped that, as you read about the many aspects of the issues explored in this book, you will critically evaluate the information presented.

It is important that you decide whether you are being presented with facts or opinions. Does the writer give a biased or unbiased report? If an opinion is being expressed, do you agree with the writer? Is there potential bias to the 'facts' or statistics behind an article?

ASSIGNMENTS

In the back of this book, you will find a selection of assignments designed to help you engage with the articles you have been reading and to explore your own opinions. Some tasks will take longer than others and there is a mixture of design, writing and research-based activities that you can complete alone or in a group.

FURTHER RESEARCH

At the end of each article we have listed its source and a website that you can visit if you would like to conduct your own research. Please remember to critically evaluate any sources that you consult and consider whether the information you are viewing is accurate and unbiased.

Useful weblinks

www.anglia.ac.uk Anglia Ruskin Universiry

www.carbonbrief.org

www.energybillrevolution.org

www.energysavingsecrets.co.uk

www.energysavingtrust.org.uk

www.foeeurope.org Friends of the Earth Europe

www.frack-off.org.uk

www.fwi.co.uk Farmers Weekly

www.gov.uk/government/organisations/department-of-energy-climate-change

www.govtoday.co.uk

www.manchester.ac.uk The University of Manchester

www.nationalgrid.com

www.nationalgridconnecting.com

www.niauk.org Nuclear Industry Association

www.ons.gov.uk Office for National Statistics

www.rt.com

www.theconversation.com

www.uswitch.com

Renewable energy vs non-renewable energy

By Kelly Fenn

The chief contributor to climate change is carbon emissions from fossil fuels. However, with those finite fuel sources running out, considering new forms of energy won't just be good for the environment, it will be absolutely vital in years to come. For now, a commitment to renewable energy is on the world political agenda, with Europe, for example, proposing a target of increasing total use of renewable energy from 7% to 20% by 2020.

Take a look into how each source of energy, renewable and non-renewable, compare.

Coal

Coal has the most widely distributed reserves in the world and is mined in over 100 countries. While scientists believe there are still adequate reserves of coal to serve the world's energy needs for some years to come, the impact of burning coal is environmentally devastating.

Burning coal is a leading cause of smog, acid rain and toxic substances in the air, and one of the chief culprits of carbon dioxide emissions. In an average year, a typical coal power station generates 3,700,000 tons of carbon dioxide, and is the primary human cause of global warming – that's as much carbon dioxide as cutting down 161 million trees.

Oil

Oil and petroleum products supply a third of the primary energy used in the UK. One of the chief problems with reliance on oil is the difficulty and cost involved in drilling for and gathering it – meaning the cost of wholesale oil is continuing to rise. Oil spillages during transportation also pose serious risks to the environment and wildlife. Additionally, burning oil also has a grave environmental impact.

Oil is the most commonly reported cause of water pollution, with over 5,000 incidents recorded annually by the Environment Agency. A single litre of oil spilled can contaminate a million litres of drinking water. And overall, 30% of CO_2 emissions affecting the atmosphere come from cars and other petrol guzzling vehicles.

Gas

Natural gas burns cleaner that the other fossil fuels and produces less greenhouse gases when processed. For an equivalent amount of heat, burning natural gas produces about 30% less carbon dioxide than burning petroleum and about 45% less than burning coal.

Before natural gas can be used as a fuel, it undergoes extensive processing to remove other materials contained in the gas, meaning that other gases escape into the air. Crucially, scientists suggest that reserves of gas will have been exhausted by 2085.

Nuclear

Nuclear energy is energy released from the atomic nucleus. Nuclear is a clean form of energy in that it releases almost none of the CO_2 emissions associated with fossil fuels.

While the amount of energy that can be produced through nuclear power is significant, so too are the possible side effects. Disposing of radioactive nuclear waste is a serious issue to consider. Nuclear energy is a controversial source of energy, with the effects caused by nuclear spills such as the Chernobyl disaster having long-term and detrimental effects on the environment and human health.

Wind

Wind farms are one of the fastest growing green sources of electricity generation. With the UK possessing 40% of Europe's total wind resource, wind power offers a compelling alternative to fossil fuel power, and it's also renewable and emission free.

In the future, if wind energy were to provide a significantly larger chunk of global energy, then the land needed for wind farms would have a big impact on more and more people's living space. There are also concerns that wind farms offshore interrupt ecosystems and local wildlife. Another complaint is that wind farms are ugly and noisy.

Hydro

Tidal and river water flow can be harnessed to create energy, and with the UK being surrounded by water, appears to be a viable energy option. River hydro stations aren't considered to be a serious option due to the ecological impact on the local environment. However, tidal hydro-power stations are growing in technological capabilities, with research and trials continuing. It's not yet as developed as other renewable technologies but certainly one to watch.

Solar

Transforming naturally occurring light rays into energy is a logical and sustainable source of energy. In spite of the short summers and cloudy skies, the UK still receives 50% of the sunlight per square foot as countries on the equator. Solar energy is popular and working well in other European countries, and could serve the UK similarly.

However, there are limitations to the technology that will need improvement so the process is optimised and more efficient – for example, the photovoltaic cells used in the process of solar harnessing only currently absorb around 15% of the sunlight's energy.

1 July 2014

⇨ The above information has been reprinted with kind permission from EnergySavingSecrets. Please visit www.energysavingsecrets.co.uk for further information.

Energy Act

The Energy Act received Royal Assent on 18 December 2013.

On Thursday, 29 November 2012, the Secretary of State for Energy and Climate Change confirmed the Introduction of the Energy Bill to the House of Commons alongside the Annual Energy Statement.

The Energy Act received Royal Assent on 18 December 2013. Details of the Bill's Parliamentary passage are available on the Parliament website.

This series brings together all of the department's documentation for the Energy Act.

The Energy and Climate Change Committee conducted pre-legislative scrutiny on the Electricity Market Reform provisions of the draft Bill. Their report was published on 23 July and can be viewed in full on the parliament.uk: Energy and Climate Change Committee web pages. The Government's response to this was published alongside the Bill's Introduction to Parliament.

The Lords informal scrutiny group, chaired by Lord Oxburgh conducted pre-legislative scrutiny of the Electricity Market Reform provisions of the draft Energy Bill. The group wrote to Lord Marland, Baroness Verma's predecessor, commenting on the draft Bill.

Baroness Verma wrote to the Lords informal scrutiny group upon the Bill's Introduction into the House of Commons on 29 November 2012, to respond to their comments.

This Act will establish a legislative framework for delivering secure, affordable and low carbon energy and includes provisions on:

Decarbonisation

These provisions enable the Secretary of State to set a 2030 decarbonisation target range for the electricity sector in secondary legislation. A decision to exercise this power will be taken once the Committee on Climate Change has provided advice on the level of the 5th Carbon Budget, which covers the corresponding period (2028–32), and when the Government has set this budget, which is due to take place in 2016.

Electricity Market Reform (EMR)

This Act puts in place measures to attract the £110 billion investment which is needed to replace current generating capacity and upgrade the grid by 2020, and to cope with a rising demand for electricity. This includes provisions for:

⇨ Contracts for Difference (CFD): long-term contracts to provide stable and predictable incentives for companies to invest in low-carbon generation;

⇨ Capacity Market: to ensure the security of electricity supply including provisions to allow Electricity Demand Reduction to be delivered;

⇨ Conflicts of Interest and Contingency Arrangements: to ensure the institution which will deliver these schemes is fit for purpose;

⇨ Investment Contracts: long-term contracts to enable early investment in advance of the CFD regime coming into force in 2014;

⇨ Access to Markets: this includes Power Purchase Agreements (PPAs), to ensure the availability of long-term contracts for independent renewable generators, and liquidity measures to enable the Government to take action to improve the liquidity of the electricity market, should it prove necessary;

⇨ Renewables Transitional: transition arrangements for investments under the

Renewables Obligation scheme; and

⇨ Emissions Performance Standard (EPS): to limit carbon dioxide emissions from new fossil fuel power stations.

Nuclear regulation

The Act places the interim Office for Nuclear Regulation (ONR) on a statutory footing as the body to regulate the safety and security of the next generation of nuclear power plants. This includes setting out the ONR's purposes and functions.

Government pipe-line and storage system

The Act includes provisions to enable the sale of the Government Pipe-line and Storage System (GPSS). This includes providing for the rights of the Secretary of State in relation to the GPSS, registration of those rights, compensation in respect of the creation of new rights or their exercise, and for transferral of ownership, as well as powers to dissolve the Oil and Pipelines Agency by order.

Strategy and policy statement

The Act improves regulatory certainty by ensuring that the Government and Ofgem are aligned at a strategic level through a Strategy and Policy Statement (SPS), as recommended in the Ofgem Review of July 2011.

Consumer protection

⇨ To set a limit on the number of energy tariffs offered to domestic consumers; require the automatic move of customers from poor value closed tariffs to cheaper deals; require the provision of information by suppliers to consumers on the best alternative deals available to them from them.

⇨ To allow Ofgem to extend its licence regime to third-party intermediaries, such as switching websites.

⇨ To provide a new enforcement power for Ofgem to require energy businesses that breach gas or electricity licence conditions, or other relevant regulatory requirements, to provide redress to consumers who suffer detriment as a result of the breach.

⇨ To amend the Warm Homes and Energy Conservation Act to propose a new target for fuel poverty that would be set through secondary legislation.

In addition:

⇨ a provision to allow the FITs scheme to include community energy projects with a capacity of up to 10 MW (an increase on the current ceiling of 5 MW).

⇨ to provide an exception to the prohibition of participating in the transmission of electricity, during testing in the commissioning period of Offshore Transmission connections, constructed by or on behalf of developers also constructing an offshore generating station.

⇨ a provision that enables DECC to charge fees for providing energy resilience services in the event of a disruption, or threatened disruption to energy supplies.

⇨ to extend the existing regime for recovering from industry the Department's external adviser costs relating to the Waste Transfer Contract and any agreement reached under Section 46 of the Energy Act 2008. This will also extend the existing cost recovery regime to cover the period between notification by an operator of its intention to submit a Funded Decommissioning Programme and formal submission.

⇨ a provision to enable the Secretary of State to require private landlords to provide smoke and/or carbon monoxide alarms.

Last updated: 18 December 2013

⇨ The above information has been reprinted with kind permission from the Department of Energy & Climate Change. Please visit www.gov.uk for further information.

Digest of UK Energy Statistics 2014

The Department of Energy & Climate Change today releases four key publications: *the* **Digest of United Kingdom Energy Statistics 2014, UK Energy in Brief, Energy Flow Chart,** *and* **Energy Consumption in the United Kingdom** *providing detailed analysis of production, transformation and consumption of energy in 2013.*

Key points

⇨ Primary energy production fell by 6.3 per cent, on a year earlier, due to record low coal output following mine closures; oil and gas output were also down as output facilities were affected by maintenance issues alongside longer term decline.

⇨ Final energy consumption rose by 0.7 per cent, reflecting the colder weather in 2013. On a temperature adjusted basis, energy consumption was down 0.3 per cent continuing the downward trend of the last nine years.

⇨ Electricity generated from renewable sources in the UK in 2013 increased by 30 per cent on a year earlier, and accounted for 14.9 per cent of total UK electricity generation, up from 11.3 per cent in 2012. Total renewables, as measured by the 2009 EU Renewables Directive, accounted for 5.2 per cent of energy consumption in 2013, up from 4.2 per cent in 2012.

⇨ In 2013, the UK became a net importer of petroleum products for the first time since 1984 (the year of the miner's strike) and before that 1973; largely due to the closure of the Coryton Refinery in July 2012.

Main energy production and trade statistics:

⇨ Primary energy production fell by 6.3 per cent in 2013, following the record falls of 13.4 and 10.8 per cent in the previous two years. Production has now fallen in each year since 1999, and is now less than half its 1999 levels, an average annual rate of decline of 6.6 per cent.

⇨ Gross natural gas production fell 6.2 per cent in 2013. This reflects the continuing long-term decline in UK natural gas production, which has fallen by an average of 8.0 per cent per year since 2000, when production peaked.

⇨ Crude oil (including NGLs) production in 2013 was 8.8 per cent lower than in 2012 at 41 million tonnes. Production has fallen by 70 per cent from its 1999 peak.

⇨ Coal production was down by 25 per cent in 2013 compared to 2012, following the closure of a number of mines.

⇨ Energy imports were at record levels in 2013, up 2.3 per cent on 2012 levels

- For crude oil, the key source was Norway which accounted for 40 per cent of imports, with a large growth in imports from Algeria and Saudi Arabia.

- For gas the key source was also Norway, which accounted for 58 per cent of UK imports, with 16 per cent from The Netherlands. LNG accounted for 20 per cent of gas imports, down from 28 per cent in 2012, with 93 per cent of these imports from Qatar.

- The UK sources its petroleum products widely, with a range of European countries supplying diesel road fuel. Aviation fuel is also sourced widely with significant volumes from OPEC countries such as Kuwait and Saudi Arabia. The UK, though, remains a net exporter of petrol with 35 per cent of exports shipped to the USA.

- For coal the key source was Russia, accounting for 41 per cent of UK imports, followed by the USA and Colombia which accounted for 25 and 23 per cent, respectively.

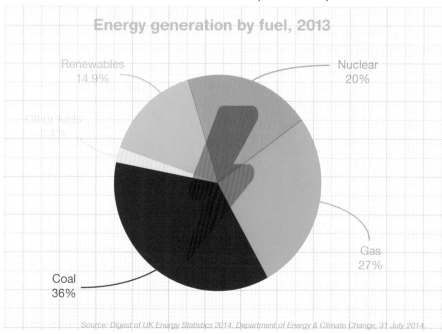

Energy generation by fuel, 2013

Renewables 14.9%

Nuclear 20%

Other fuels 2.1%

Coal 36%

Gas 27%

Source: Digest of UK Energy Statistics 2014, Department of Energy & Climate Change, 31 July 2014.

- The UK remained a net importer of energy, with an increased dependency level (imports/energy use) of 47 per cent; this continues the trend from 2004 when the UK once again became a net importer of fuel. In 2013 the UK was a net importer of all fuels, as imports of petroleum products in total exceeded exports following the closure of the Coryton refinery.

Main energy consumption statistics:

- UK primary energy consumption in 2013 decreased by 0.6 per cent, and on a temperature-adjusted basis, consumption was down 1.9 per cent, continuing the downward trend of the last eight years.

- Overall gas demand decreased by 1.1 per cent. Gas demand for electricity generation decreased by 5.7 per cent as gas's share of the UK's generation of electricity fell to 27 per cent, from 28 per cent last year. Domestic demand was similar to that in 2012.

- Total oil consumption in the UK saw a small fall down by one per cent when compared with 2012. Over 70 per cent of oil is consumed in the transport sector, which showed little change in overall consumption from 2012.

- Consumption of diesel road fuel exceeded the consumption of motor spirit in 2013 by over nine million tonnes. Up until 2005, motor spirit exceeded diesel road fuel sales, since then a large element of the UK's car fleet has switched to diesel. Petrol consumption has fallen by around four per cent per annum in the past ten years, whilst diesel use has increased by nearly two per cent per annum, over the same period.

- Coal consumption decreased by 5.7 per cent in 2013. There

It's a grand day for wind power...

I suppose you're right!

was a 7.4 per cent decrease in consumption by major power producers (consumers of 83 per cent of total coal demand) reflecting lower demand and more renewables. Coal accounted for 36 per cent of the electricity generated in the UK in 2013, down from 39 per cent in 2012. The domestic sector accounted for only 1.1 per cent of total coal consumption.

- Energy consumption by final users at 142.5 million tonnes of oil equivalent increased by 0.7 per cent in 2013. Consumption in the domestic sector was broadly unchanged, up only 0.2 per cent; with industry and service sector use up by 2.4 and 3.1 per cent, respectively. There was reduced consumption from transport, which was down 0.7 per cent. Average temperatures in 2013 were marginally cooler than in 2012. On a temperature-adjusted basis final energy consumption was down by 0.3 per cent continuing the downward trend of the last nine years.

- Refinery production decreased by six per cent on 2012 and 25 per cent on 2000. The closure of the Coryton refinery contributed to the decrease in production as it had been operating in the first half of 2012. Production was

dented further by the temporary closure of the Grangemouth refinery in October 2013. Imports of petroleum product imports have increased by eight per cent to make up the shortfall. In addition, exports have decreased by three per cent, as a result the UK was a net importer of petroleum products in 2013 for the first time since 1984, the year of the miners' strike. Petrol accounts for over a third of exports.

Main electricity generation and supply statistics:

- Gas prices remained high in 2013, such that the commercial attractiveness of gas for electricity generation continued to be weak in 2013. Meanwhile, nuclear's share of electricity generation was unchanged, despite a slight increase in generation. Gas accounted for 27 per cent of electricity supplied in 2013, with coal accounting for 36 per cent and nuclear 20 per cent.

- Electricity generated from renewable sources in the UK in 2013 increased by 30 per cent on a year earlier, and accounted for 14.9 per cent of total UK electricity generation, up from 11.3 per cent in 2012. Offshore wind generation increased by 50 per cent, and onshore wind increased by 40 per cent. Both the offshore and onshore wind load factors (37.5 per cent and 27.9 per cent) exceeded or equalled that of gas (27.9 per cent).

⇨ In 2013, the proportion of UK electricity generated from renewables was 14.9 per cent. Installed electrical generating capacity of renewable sources rose by 27 per cent (4.2 GW) in 2013, mainly as a result of a 27 per cent increase (1.6 GW) in onshore wind capacity, 59 per cent increase (1.0 GW) in solar photovoltaic capacity (due to high deployment of both small-scale capacity under Feed in Tariffs and large-scale capacity under the Renewables Obligation). Bioenergy capacity increased by 27 per cent (0.8 GW) due to new conversions of previously coal-fired capacity to biomass.

⇨ There was a 0.5 per cent decrease in the total supply of electricity in the UK in 2013, to 373.6 TWh. Indigenous electricity supply fell by one per cent, but net imports of electricity increased by around 20 per cent, to 14.5 TWh, as imports rose substantially more than exports.

⇨ Final consumption of electricity was broadly unchanged at 317.3 TWh, the lowest level since 1998.

⇨ The domestic sector was the largest electricity consumer in 2013 (113.5 TWh), while the industrial sector consumed 98.0 TWh, and the service sector consumed 101.7 TWh. Industrial consumption increased by 0.2 per cent, while domestic consumption fell by 1.1 per cent.

Other energy statistics:

⇨ Total renewables, as measured by the 2009 EU Renewables Directive, accounted for 5.2 per cent of energy consumption in 2013 up from 4.2 per cent in 2012.

⇨ In 2013, Combined Heat and Power (CHP) capacity stood at 6,170 MWe, a small decrease on 2012.

⇨ In 2013 the energy industries accounted for 3.3 per cent of GDP.

The fuel switching away from gas and coal for electricity generation, with other changes, is provisionally estimated to have decreased emissions of carbon dioxide by around two per cent in 2013.

Energy consumption in the United Kingdom

⇨ Final energy consumption, excluding non-energy use, rose by 0.9 million tonnes of oil equivalent (mtoe) from 141.5 to 142.5 mtoe between 2012 and 2013 – an increase of one per cent.

⇨ Energy consumption in 2013 was 16.9 mtoe lower than in 2000 (142.5 mtoe compared to 159.4 mtoe) – a decrease of 11 per cent, and two per cent lower than in 1970.

⇨ In 2013, energy consumption in the industrial sector increased by two per cent since 2012, with the iron and steel sector showing a 17 per cent increase to 1.3 mtoe. The largest energy consuming single sub-sector in the industrial sector was chemicals, which accounted for 14 per cent of all industrial energy consumption. Energy consumption per unit of output fell by 53 per cent in the chemicals sector between 2000 and 2013, while there was a fall of eight per cent in the same measure for the iron and steel sector; for all industries there was a fall of 19 per cent.

⇨ Energy consumption in the transport sector decreased by 0.7 per cent between 2012 and 2013. Transport energy consumption fell four per cent (2.0 mtoe) between 2000 and 2013, with the largest actual decrease occurring in the road transport sector, where consumption fell by four per cent (18 mtoe) – with this sector accounting for 74 per cent of total transport consumption in 2013. Over the same period, air transport fuel increased by two per cent since 2000 and rail transport use fell by 23 per cent.

⇨ In 2013, domestic energy consumption remained stable with consumption in 2012 (0.2 per cent increase) – similar levels of consumption to 2009. The intermediate years had unusual weather spells (the high levels of consumption in 2010 were largely driven by colder temperatures and the lower levels of consumption in 2011 were due to a warmer than usual heating season).

⇨ The seven per cent decrease in consumption since 2000 is set in the context of an increase of 11 per cent in the number of UK households and a nine per cent increase in the UK population. At a per household level, energy consumption has fallen by nine per cent since 2000.

⇨ In the service sector, energy consumption in the private commercial sector increased by 30 per cent between 2000 and 2013, in the public sector it fell by 23 per cent and by 23 per cent in the agriculture sector. Over the same period, output, measured as the contribution made to the UK economy, increased by 35 per cent in the private sector and increased by 23 per cent in the public sector, in real terms.

31 July 2014

⇨ The above information has been reprinted with kind permission from the Department of Energy & Climate Change. Please visit www.gov.uk for further information.

Household energy spending in the UK, 2002–2012

Household energy spending: seven things you should know.

1. UK households spent an average of £106 a month on household energy in 2012. This was a 55% rise on the 2002 monthly spend, after accounting for inflation. This is despite a decline in average energy usage.

UK households spent an average of £106 a month on electricity, gas and other household fuels[1] in 2012. This compares to £69 a month in 2002, after adjusting for inflation using the Consumer Price Index (CPI). This means there was a 55% increase in average household spending on energy between 2002 and 2012.

Over this same time period, household energy use has fallen, with figures from the Department of Energy and Climate Change (DECC) showing that the average amount of energy used per household was 17% lower in 2012 than in 2002. This means the increase in the average amount households are spending is explained solely by rises in energy prices.

Average household energy use has fallen for a number of reasons. These include the installation of energy efficiency measures (such as loft and wall insulation, and more efficient boilers), households responding to higher bills by reducing use, and generally increasing public awareness of energy consumption and environmental issues. However, energy prices have risen faster than average household consumption has fallen, resulting in rising average bills.

Most of the increase in average spending on energy came between 2004 and 2009, reflecting the significant increases in energy prices which occurred over this period. By 2009, the average household spent the equivalent of £108 a month (in 2012 prices) on energy. Since then, there has been relatively little change in the average monthly spend on energy.

The limited change in expenditure since 2009 may be explained by a combination of factors, including a fall in domestic energy prices in 2010, and price rises in subsequent years appearing to be offset by lower energy use, partly due to milder winter temperatures.

2. Average household spending on gas increased 56%, while average spending on electricity increased 43% between 2002 and 2012, after accounting for inflation.

In 2002, households using electricity had an average monthly spend of £35 (in 2012 prices). In 2012 this had risen to £51, a 43% increase. Households that used gas in 2002 were spending an average of £37 a month on this fuel (in 2012 prices)

but this had risen to £57 in 2012, a 56% rise.

The larger increase in the average spend on gas is largely explained by the fact that gas prices have risen more sharply than electricity prices in recent years, rather than any changes in the amounts used. Despite some year-on-year fluctuations, possibly reflecting variations in average temperatures, there have not been any substantial changes in the fuel mix for household use over this time. Figures from DECC show that in 2012, 67.6% of household energy use was natural gas, while electricity made up 22.9%. This represents only a relatively small change from 2002 when the percentages were 68.2% and 21.7%, respectively.

In 2012, only 7% of households recorded spending on other household fuels such as coal, oil for central heating, paraffin and wood. However, for these households,

Average monthly spend on energy for households across the UK: 2010–2012

£112

£154

£103

Households in Northern Ireland have a substantially higher average energy spend than households in the rest of the UK

£105

Source: Household Energy Spending: 7 things you should know, *Office for National Statistics, 3 March 2014.*

1   These include coal, oil for central heating, paraffin and wood. Transport fuels (i.e. petrol and diesel) are not covered by this article.

average spend on these fuels was £131 a month in 2012, up from £68 in 2002, a 92% increase.

3. On average, households spent the equivalent of 5.1% of their income on household energy in 2012, up from 3.3% in 2002. Most of this rise occurred between 2004 and 2009.

While spending on energy has risen significantly between 2002 and 2012, households have not seen a comparable increase in their disposable income, that is, the amount of money that they have available for spending and saving after accounting for direct taxes (such as income tax and council tax). Between 2002 and 2007, the average household's disposable income grew by 6.9%, after taking into account the effect of inflation. However, since 2007 disposable incomes have fallen by approximately 6.7% in real terms.

As a result of this, expenditure on household energy is now equivalent to 5.1% of household disposable income for the average household, up from 3.3% in 2002. Most of this increase occurred between 2004 and 2009, although the fall in disposable income since 2007 has increased the impact of recent energy price rises.

4. The poorest fifth of households spent 11% of their income on household energy in 2012, up from 8% in 2002. The richest fifth spent just 3% in 2012 up from 2% in 2002.

In 2012, the poorest fifth of households spent £93 a month on household energy (equivalent to 11% of the average disposable income for this group), compared with £126 a month for the richest fifth (equivalent to 3% of their disposable income). This difference is likely to be partly explained by the poorest fifth of households having, on average, fewer rooms than the richest fifth, which may influence their energy needs and therefore use. However, it is probable that a range of other factors which affect spending decisions and priorities are also important.

Concentrating on the poorest fifth of households, the percentage of their disposable income spent on

Heads we turn the heater on...

...tails we go cold but can afford to eat dinner.

household energy has risen from 8% in 2002 to 11% in 2012. This increase is explained by energy prices rising at a faster rate than incomes. Although, after adjusting for inflation, disposable incomes for the poorest fifth of households grew by around 11% between 2002 and 2012, average spend on energy rose by 51%.

As with the overall population, the majority of this increase occurred between 2004 and 2009, reflecting the significant price increases over this period. The slight fall in the percentage of income spent on energy in 2011 may be due to a combination of factors, including a reduction in spending resulting from a relatively mild winter compared to the year before.

For the richest fifth of households, spending on energy grew by 36% between 2002 and 2012, while average disposable incomes actually fell by 1% over this period, after adjusting for inflation.

5. Retired households consistently spent a greater percentage of their income on household fuel than non-retired households, even after accounting for winter fuel payments.

In 2012, retired households spent, on average, £97 a month on household energy, compared with £110 for non-

retired households. However, when taken as a percentage of disposable income (which includes Winter Fuel and Cold Weather payments, as well as other cash benefits, pensions, and earnings from employment), spending on energy is consistently higher for retired households than non-retired households. In 2012, retired households spent an average of 7% of their disposable income on household energy, compared with 4% for non-retired households.

The difference in the amount spent by retired and non-retired households is likely to be partly due to retired households having fewer people in them, on average, resulting in lower total energy requirements (though a higher energy requirement per person). In 2012, the average number of people in a retired household was 1.5 and in a non-retired household it was 2.6.

All figures on household spending on energy in this article are based on the amount actually paid, so are net of any discounts offered by suppliers, including those provided through the Warm Homes Discount Scheme. The Warm Homes Discount, which was introduced in 2011, is offered to people whose electricity supplier is part of the scheme and either receive the Guarantee element of Pension Credit or meet 'broader group' criteria, which can vary across suppliers.

6. The average Winter Fuel/Cold Weather payment to a retired household was equivalent to 20% of their typical energy bill in 2012.

In 2012, retired households received an average of £216 a year in Winter Fuel and Cold Weather Payments. This was equivalent to around 20% of their typical annual energy bill. Non-retired households received an average of £24 a year in these payments, equating to around 2% of their annual spending on household energy.

Winter Fuel Payments of between £100 and £300 a year are paid to UK residents over the qualifying age (for Winter 2013/14, it is those born on or before 5 January 1952). Additionally, people who are in receipt of certain benefits such as Pension Credit and Income Support may be eligible for Cold Weather Payments during periods of prolonged cold weather.

7. The average household in Northern Ireland spent a substantially higher amount on energy than the average household in other UK countries.

Considering the average household energy spend across the countries of the UK from 2010 to 2012, households in Northern Ireland were spending £154 a month on energy (in 2012 prices). This was £42 higher than in Scotland, £49 higher than in Wales and £51 higher than in England.

The higher average spend in Northern Ireland may be partly explained by the different nature and size of the energy market there. Due to a relatively limited gas network, many homes in Northern Ireland rely on oil-fired central heating, which is likely to lead to higher energy costs (the Northern Ireland House Conditions survey found that 68% of households used heating oil for central heating). Additionally, in those areas where mains gas is available, the market is not yet fully open to competition, with only a single supplier operating in some areas.

Across the rest of the UK, any differences in the average energy spend are very small. The average monthly spend in Scotland is slightly higher than in England and Wales, consistent with DECC figures on energy consumption per household.

3 March 2014

⇨ The above information has been reprinted with kind permission from the Office for National Statistics. Please visit www.ons. gov.uk for further information.

Energy prices rise forcing changes in lifestyle

By Ben Tobin

A recent YouGov Reports publication has found that three in ten are spending 10% or more of their household income on gas and electricity, as the debate surrounding energy prices rages on in the Commons and beyond.

Of those in this group, many say they have had to make changes to their lives in order to pay the bills. Almost seven in ten (68%) say they have turned the heating down or off when they ordinarily would have left it on, 27% have spent less on food while 5% have borrowed from short term lenders in order to fulfil bills. 44% of those not in this group say they have had to reduce their usage.

Negative opinions towards energy suppliers and prices are commonplace. 84% agree that companies are quick to raise prices when their costs go up, but slower in offering discounts when they fall. Over two thirds (67%) agree that big energy suppliers act as a cartel, while 66% say the electricity and gas supply market has major problems which the Government needs to address. Half (50%) believe UK household energy bills are some of the highest in Europe. Only 9% say big suppliers treat their customers fairly.

Changes to the market

In terms of changes to the market, 69% favour forcing energy supplies to reduce the number of different tariffs they offer and simplify bills. Over half (52%) are positive towards the use of one-off windfall taxes on profits of large energy supplies to help cut household bills, or to support infrastructure needs. 51% support freezing household energy prices for 20 months from May 2015 and 46% would like to see greater regulation in order to force greater competition.

Tom Rees, UK Research Manager at YouGov Reports, said; 'Our research indicates how little consumers trust energy suppliers, the high level of dissatisfaction with the energy market and the tangible effect on household finances. Whether this will lead to consumers switching providers in greater numbers remains to be seen, and the question of what the Government should do will be crucial in the lead up to next May's election.'

19 June 2014

⇨ The above information has been reprinted with kind permission from YouGov. Please visit www. yougov.co.uk for further information.

Half of UK householders think it's cheaper to leave the home heating on all day at low temperature

Energy Saving Trust launches energy saving mythbusters in bid to help UK householders cut fuel bills.

Half (50 per cent) of all UK householders think it's cheaper to leave their home heating on all day than turning the heating on or off and up or down when required.

This finding is part of an energy saving mythbusting survey commissioned by the Energy Saving Trust as part of Big Energy Saving Week, which highlights the perceptions of the UK public and how they don't always match the reality of the energy saving action or statement.

The survey was carried out to a sample of 2,067 online adults in the UK aged 16–75 earlier this month.

Other findings from the Ipsos MORI survey include:

⇨ 30 per cent that think screensavers on computers save energy.

⇨ Almost half (45 per cent) think switched-off electrical appliances don't use electricity when they're plugged in at the mains.

⇨ Nearly half (48 per cent) agreed that it's a hassle to change energy suppliers.

⇨ Two thirds (66 per cent) of people think that more heat is lost through the roof of their home than the walls – however, for the majority of properties the walls will actually lose more heat.

However, there were other findings which highlighted a level of awareness from the UK public regarding the correct energy saving actions to take in the home. These include:

⇨ Just 27 per cent incorrectly said turning up their thermostat to a high setting heats the home faster, compared to 54 per cent that correctly said the statement was false.

⇨ Three quarters (75 per cent) know that energy saving light bulbs are compatible with traditional light bulb fittings.

⇨ And, 69 per cent know that solar panels will work during daylight hours regardless of whether the sun is shining or not, with only ten per cent disagreeing.

Energy Saving Trust chief executive Phillip Sellwood said: 'We commissioned this survey to bust some of the top energy saving myths we encounter on a daily basis. While for certain actions a portion of the UK public think they are saving energy when they're not, it's heartening to see that a lot of people are doing the right thing in the home to save energy.

'We know it's important for the UK public to stay warm and cosy in their homes. But for the majority the most cost-effective way to do this is to turn the heating on and off or up and down when required rather than leaving it on all day at a lower temperature. This ensures that heat is not wasted and that your home will be at a comfortable temperature.

'Our message to UK householders is clear: if you don't know where to start when it comes to saving energy and cutting your fuel bill, or just need some advice, visit the Energy Saving Trust website or give the Energy Saving Advice Service a call on 0300 123 1234 for independent and impartial advice.

'Throughout Big Energy Saving Week, the Energy Saving Trust and partners will provide practical solutions and advice to help households save money on their energy bills. For us, saving energy – not wasting it – is the right thing to do and makes total sense.'

According to figures from the Energy Saving Trust, the UK could collectively save nearly £4.4 billion on energy bills if householders took three energy saving actions in the home.

These are:

⇨ **Turn it off** – Make sure you turn your lights, appliances and chargers off when you're not using them. Virtually all electrical and electronic appliances can safely be turned off at the plug without upsetting their systems.

⇨ **Turn it down** – Many households have their central heating set higher than they need, without even realising it. If it's too warm inside, try turning your room thermostat down by one degree and see if you are still at a comfortable temperature. Every degree that you turn it down will make additional savings to your heating bill.

⇨ **Let there be light** – Households can now get LED spotlights that are bright enough to replace halogens, as well as regular energy saving bulbs ('compact fluorescent lamps' or CFLs) for pretty much everything else. They come in a variety of shapes, sizes and fittings and can save households money on their energy bills.

Big Energy Saving Week took place between Monday 27 and Friday 31 January, with the week aiming to raise awareness of energy and efficiency issues among the UK public through joint working between the voluntary sector and energy suppliers.

The week has taken place twice previously and is funded by the largest six energy companies., with involvement from Citizens Advice Bureau, Energy Saving Trust, Age UK, ACRE (Action with Communities in Rural England), Consumer Futures, the Department of Energy and Climate Change (DECC), Ofgem, Energy UK and National Energy Action.

To find out more visit www. bigenergysavingweek.org.uk or call 0300 123 1234.

Ten energy saving myths

1. Leaving the heating on all day on a low temperature is cheaper than turning the heating up and down or on and off as needed.

FALSE: For the majority of householders leaving your room thermostat on all day at a lower temperature will not only mean that your home will never be at a comfortable temperature but it will also waste heat when you do not need it. Room thermostats turn the heat on and off when your home reaches the set temperature that you feel comfortable at. Combine this with a timer control that tells your heating system to come on only when you need it to save money on your energy bills.

2. Cranking up the thermostat heats your home faster.

FALSE: Your room thermostat turns your heating system on or off according to a set temperature. No matter how high you set the temperature, the rate at which your central heating distributes heat remains constant. To heat your home faster, install better insulation. This decreases the rate at which heat is lost through your walls, loft, windows and floor – heating your home faster and keeping it warm for longer.

3. Electrical appliances, such as TVs, laptops, phone chargers, etc., don't use electricity when they're plugged-in but not in-use.

FALSE: Some electrical appliances and chargers draw energy even when the devices are not being used. This 'vampire power' wastes energy, and the best way to avoid this is to remember to switch off at the wall and pull out the plug.

By avoiding standby, and making sure devices are not left plugged in or idle, a typical home could save between £50–£80 a year.*

*Based all home appliances, consumer electronics, lights and chargers that have been left on standby mode or have been left on and not in use, using the average electricity cost of 13.52p/kWh.

Sourced from DEFRA's Home Electricity Study.

4. It is cheaper to run appliances, such as washing machines, at night than during the day.

This may be true, but not for most of us. While some households in the UK are on tariffs that vary depending on the time of day, such as Economy 7, the majority of customers pay the same rate at all times of day and night. However, if you know you are already on an Economy tariff, or are considering switching to one, then running appliances during off-peak periods will be cheaper.

5. With traditional light-bulb fittings, you cannot do a straightforward swap with energy saving bulbs and LED light bulbs.

FALSE: Energy saving and LED light bulbs come in all shapes and sizes and can now be fitted in down-lighters, free-standing lamps and traditional pendants.

6. Putting plastic tape and a layer of cling-film around draughty windows is a better option at keeping heat in the home than draught excluders or double glazing.

FALSE: Although physically blocking gaps around your windows with cling film or plastic tape may stop draughts and reduce heat loss, this will not be as effective as draught excluders or double glazing. These more permanent measures reduce heat loss more effectively – keeping your home warmer and saving money on your heating bills.

7. Cavity wall insulation causes damp in the home.

FALSE: In most cases cavity wall insulation is likely to alleviate and not exacerbate damp in a home. A combination of proper insulation, adequate ventilation and balanced heating in a home will help avoid cold spots and moisture from condensing on your walls. Assessors should be able to advise you as to whether your home is suitable for insulation and any potential risk from damp.

8. Solar panels don't generate electricity on a cloudy day.

FALSE: Whilst solar panels will work most effectively in bright sunlight, they nonetheless continue to collect energy from diffuse light even on a cloudy day. Summer months are the most productive as there are longer daylight hours than in winter.

9. When using a desktop computer, screensavers save energy.

FALSE: Because your screen remains on, screensavers are basically another program which consumes energy like any other. While computers have timed sleep settings which do use less energy, switching off your monitor or even your whole computer when taking breaks is the most effective way to stop energy being wasted.

On average desktop computers cost around £24 a year to run.*

*Based all home appliances, consumer electronics, lights and chargers that have been left on standby mode or have been left on and not in use, using the average electricity cost of 13.52p/kWh. Sourced from DEFRA's Home Electricity Study.

10. It's difficult and a hassle to switch energy suppliers.

FALSE: There are a number of energy price comparison companies where you can find the cheapest tariff for your area by checking online or by telephone. Once you have filled in the application with your main details, which typically takes around half an hour, the energy provider will sort the switch for you. Finding the deal that's best for you, and switching energy supplier, can be a great way to reduce your energy bills.

⇨ The above information has been reprinted with kind permission from Energy Saving Trust. Please visit www.energysavingtrust.org.uk for further information.

© Energy Saving Trust 2015

Energy efficiency and excess winter deaths: comparing the UK and Sweden

1. Introduction

David Cameron pledged in February 2013 that he wanted the UK to become 'the most energy efficient country in Europe'[1]. However, at present the UK is the 'cold man of Europe', with very high levels of fuel poverty and poorly insulated homes. In contrast, Sweden has well-insulated homes, meaning that its people suffer less from fuel poverty and excess winter deaths – even though they face higher energy prices and colder winters.

Earlier this year, a report[2] by ACE Research for the Energy Bill Revolution compared fuel poverty and energy efficiency in the UK to 15 other European countries. The UK was ranked either the worst, or among the worst, on energy poverty, affordability of heating, efficiency of homes and excess winter deaths. This is despite the fact that it has amongst the lowest energy prices in Europe and relatively high household incomes compared to the other countries. The report showed that the poor energy efficiency of our housing stock is one of the main causes of these problems.

This briefing looks in more detail at the UK's performance relative to Sweden, which is generally seen as representing the best practice in Europe on these energy issues, and has a similar income level to the UK. Table 1[3] summarises the key findings on a range of indicators, and what these mean.

The example of Sweden provides a clear illustration of what improving the housing stock would mean for the UK – cutting fuel poverty, reducing excess winter deaths and making energy more affordable. The Energy Bill Revolution is calling for the carbon tax every household pays via their bills to be used to make UK homes highly energy efficient, prioritising the homes of the fuel poor. There is enough carbon tax revenue from the Emissions Trading Scheme and the Carbon Price Floor to end fuel poverty and significantly reduce carbon emissions and energy bills. This investment in homes is also one of the best ways to generate growth and jobs in the UK economy.

2. Background: comparing the UK and Sweden

If we are going to compare energy issues in the two countries, it is important to understand some of their basic similarities and differences.

2.1 Climate

In central and southern Sweden, summer temperatures are similar to those in England, but winters are colder and very snowy, with the sea often freezing over. In the rest of Sweden, temperatures are much colder, with long, severe winters and some areas where the snow never melts.[4] For example, comparing January temperatures in the two capital cities shows Stockholm being much colder than London (January average min. and max. daily temperature in °C: London 2–6, Stockholm -5– -1).[5]

So Sweden has much colder winters than the UK, and we might expect Swedes to have more difficulty keeping warm than Britons.

1. http://www.ukace.org/2013/02/david-camerons-speech-at-the-launch-of-deccs-energy-efficiency-mission/

2. http://www.ukace.org/2013/03/fact-file-the-cold-man-of-europe/

3. For the UK, the most common type of heating is gas heating, and for Sweden it is for district heating. To enable a fair comparison, we have converted the unit price of gas (4.5p) into a price per unit of 'useful heat'.

4. Source: The Met Office: http://www.metoffice.gov.uk/weather/europe/sweden_past.html

5. Source: http://www.bbc.co.uk/weather

Table 1: Overview of key findings for the UK and Sweden

Indicator	UK	Sweden	What does this mean?
Excess winter deaths as a proportion of all deaths (%)	4.61	3.76	High winter death rates are often linked with fuel poverty. Compared to Sweden, the UK has a 23% higher rate of excess deaths in the winter despite the fact the UK has much milder winters than Sweden.
Proportion unable to afford to heat their home adequately (%)	6.5	1.6	The share of people who cannot afford to heat their home is four times higher in the UK than in Sweden.
Proportion spending a considerable share of expenditure on energy (%)	19.2	11.2	The share of households who must spend a lot of their budget on energy is 70% higher in the UK than in Sweden.
Approximate price per unit of most common type of heating 4(p/kwh)	5.6	9	Swedes use different types of heating to Britons. However, in general, Swedes pay a higher price for each unit of energy for heating. This means prices are not the cause of the UK's poor performance on the measures above.
Real adjusted gross disposable income of households per capita (£)	18,900	19,000	The UK has almost the same household income as Sweden. This means income is unlikely to be the main cause of the UK's poor performance.
Proportion living in homes in a poor condition (%)	15.9	8.4	The UK's share of people living in leaky homes is almost twice as high as Sweden's. This is likely to be a cause of the UK's poor performance.
Average U value of walls (W/m2K)	1.16	0.35	The UK's homes lose three times more heat than Sweden's because they are poorly insulated. This is likely to be a major cause of the UK's poor performance.

2.2 Income

Another important consideration, if we are to compare the two countries, is their incomes. People in Sweden and the UK have very similar incomes. The average disposable income in the UK is £18,900. This is very slightly lower than in Sweden, where it is £19,000.[6] So there is no reason to think that differences in income would lead one country's citizens to struggle more with heating costs.

3. Results

In this section we consider the differences in fuel poverty and excess winter deaths between the two countries, and their possible explanations.

3.1 Excess winter deaths and affordable warmth

First, we investigate the issues of affordable warmth and excess winter mortality in the UK and Sweden. In a working paper in late 2010,[7] the European Commission calculated how many households in each country were spending a considerable share of their expenditure on energy (based on a predetermined threshold). Those spending over this threshold were classed as being in energy poverty. On this basis the UK has 19% of households in energy poverty,[8] while Sweden has only 11%.

Another indicator is whether people can afford to heat their homes adequately.[9] The European Commission Statistics on Income and Living Conditions (SILC) survey (2011) asked householders whether they could afford to adequately heat their home. 6.5% of UK householders said they cannot afford to keep their home warm. In contrast, in Sweden, the figure was only 1.6%.

Another important issue is excess winter mortality. This refers to the number of additional deaths in the winter months compared to the

rest of the year, and is a problem associated with fuel poverty. Table 2 shows the number of excess winter deaths was much higher in the UK in 2011–12, but this is because the UK's population is greater, and so the total number of deaths is higher. For a fairer comparison, we can look at excess winter deaths as a proportion of all deaths. This shows that in Sweden, 3.8% of all deaths were excess winter deaths, while in the UK the figure was 4.6%,[10] which is 23% higher.

Table 2: Excess winter deaths (EWDs) for 2011-2012[11]

Country	EWDs	Total deaths	Proportion of deaths that are EWDs
Sweden	3,385	89,959	3.76%
UK	25,535	554,156	4.61%

It should be noted that this method does not take account of the specific climatic conditions in the countries in that year. However, it appears that Sweden has fewer excess winter deaths as a proportion of all deaths, despite the fact that Sweden experiences far colder winters.

Across all the measures, the UK performs worse than Sweden, with more UK residents spending large amounts of their budget on energy and struggling to afford heating. The links between cold homes and negative health impacts are now well known, so this lack of affordable warmth is likely to contribute to the UK's higher rate of excess deaths in winter. This raises an important question: why are UK residents suffering more than Swedes in terms of energy poverty and its impacts?

3.2 Energy prices

It might be assumed that UK residents are struggling to keep warm because of high energy prices. Whilst it is true that retail gas and electricity prices have seen significant increases in the last few years, the UK still has relatively low energy prices compared to other European countries.

In the UK, the main source of energy for heating is gas, and DECC's *Quarterly Energy Prices* update[12] shows that Great Britain's average gas price is 4.5p per unit. To compare this to the costs of other heating sources we can convert it to a cost per unit of useful heat – this gives a cost of 5.6p per unit. Average UK domestic gas prices, including taxes, are the second lowest in the EU15. Swedes do not use gas as a main source of heating; their main sources are district heating, electricity and biomass. The average unit cost of district heating in Sweden is approximately 9p and the average unit cost of biomass is around 6–7p.[13]

DECC's *Quarterly Energy Prices* update states that in 2012, average UK domestic electricity prices, including taxes, were the fifth lowest in the EU15, at 14p per unit. DECC state that in Sweden, the figure is very slightly higher, and the Swedish Energy Agency states that the average unit cost of electric heating for a house is around 16p.[13]

These facts make the data presented here so far all the more significant – UK residents struggle more than Swedes to afford their heating bills, and have higher levels of energy poverty and excess winter deaths, despite facing lower energy prices. We already know that the two countries have similar incomes. So to understand the reasons for the observed differences between Sweden and the UK, we must compare the energy efficiency of homes in the two countries.

3.3 Quality of homes

To find out why the UK performs worse than Sweden in terms of heating affordability and energy poverty, we need to consider the state of the buildings people live in. With the available data, there are two main ways this can be measured. First, we can examine the number of householders living in a dwelling with a 'leaking roof, damp walls, floors or foundation, or rot in the window frames or floor' using the answers provided by households to the SILC survey (2011). These sub-standard homes may be hard to keep warm, and can present a health risk

---

6. We use 2011 'real adjusted gross disposable income of households per capita' from Eurostat (this is the latest data available). This is based on income before housing costs, because data on income after housing costs were not available.

7. EC, 2010

8. In fact, out of all 27 Member States, the UK ranks second from bottom according to this measure.

9. It is important to note that 'adequately' warm is a subjective measure of an expectation of comfort which undoubtedly varies from country to country. People may also have different understandings of what it means to 'afford' their heating.

10. The ONS also has a method for calculating an Excess Winter Mortality Index (EWMI) that enables comparisons of this problem across different areas. Applying this method shows that the UK in 2010/2011 had an EWMI of 14.5%. In contrast, Sweden had an EWMI of only 11.7%.

11. Data from Eurostat, 2013

12. DECC, 2013

13. Swedish Energy Agency, 2012

to occupants.[14] Sweden has 8.4% of people living in these leaky homes, while the UK has a much higher rate of 15.9%.

A second indicator of housing quality is the U-value of walls. A U-value is a measure of how much heat is lost through a building's fabric, with low values representing less heat loss (i.e. more efficient walls).

The UK's walls currently have an average U-value of 1.16 (which, for comparison, is similar to that of a modern double glazed window). In contrast, Sweden has an average of 0.35 (which is around the value of an insulated cavity wall).[15] So Swedish walls are typically over three times more efficient than British walls. Or, to put it another way, a British wall typically loses three times more heat than a Swedish wall.[16]

Together, these data suggest that the UK's buildings perform badly in terms of energy efficiency. The poor state of our buildings is a key reason why so many UK people cannot afford their heating, and are at risk of cold homes, fuel poverty, and negative impacts on their health and well-being.

4. Conclusions

This analysis shows that the share of people who cannot afford to heat their home is four times higher in the UK than in Sweden. The share of households in energy poverty is 70% higher in the UK than in Sweden. These shocking facts may explain why the UK has higher rates of excess deaths in the winter than Sweden.

This is all the more worrying when we consider the fact that the UK's energy prices are amongst the very lowest in Europe (including taxes),

with Swedes paying higher energy prices, despite having similar average household incomes. At the same time, Swedes face much colder and more severe winters.

So, if energy prices and incomes are not the main cause of the problem, we must consider the state of the housing stock. The UK's share of people living in leaky homes is almost twice as high as Sweden's, and our walls lose three times as much heat. To a considerable extent, the UK's problems of fuel poverty are explained by the inefficiency and poor state of repair of our homes.

The case of Sweden suggests that energy efficient buildings can make warmth affordable, even when energy prices are high and winters are cold. This example provides a clear illustration of what improving the housing stock would mean for the UK – cutting fuel poverty, reducing excess winter deaths and making energy more affordable for everyone.

These findings confirm that the political, social, environmental and economic opportunities available in making our housing stock one of the most efficient in Europe are very real – as already shown in work by Cambridge Econometrics and Verco (2012) and Camco (2012).

Recycling carbon revenues to make homes highly energy efficient is the best way to bring down household energy bills and the best long-term solution to end fuel poverty. It is also the most cost-effective way to reduce carbon emissions and one of the best ways for the UK to generate jobs and growth.

Bibliography

BPIE. 2013. 'Data Hub for the Energy Performance of Buildings.' *Data Hub for the Energy Performance of Buildings.* http://www.buildingsdata.eu/

Cambridge Econometrics, and Verco. 2012. *Jobs, Growth and Warmer Homes – Evaluating the Economic Stimulus of Investing in Energy Efficiency Measures in Fuel Poor Homes.* London: Consumer Focus. http://www.consumerfocus.org.uk/files/2012/11/Jobs-growth-and-warmer-homes-November-2012.pdf.

Camco. 2012. *Energy Bill Revolution Campaign Report.* London: Transform UK, The co-operative and Consumer Focus. http://www.energybillrevolution.org/wp-content/uploads/2012/02/Energy-Bill-Revolution_full-report.pdf.

Cameron, David. 2013. 'David Cameron's Speech at the Launch of DECC's Energy Efficiency Mission.' *Association for the Conservation of Energy.* http://www.ukace.org/2013/02/david-camerons-speech-at-the-launch-of-deccs-energy-efficiency-mission/.

DECC. 2013. *Quarterly Energy Prices.* London: Department of Energy & Climate Change. https://www.gov.uk/government/uploads/system/uploads/attachment_data/file/244580/qep_sep_13.pdf

EC. 2010. *Commission Staff Working Paper: An Energy Policy for Consumers.* Brussels: European Commission. http://ec.europa.eu/energy/gas_electricity/doc/forum_citizen_energy/sec(2010)1407.pdf.

Eurostat. 2013. 'European Union Statistics on Income and Living Conditions.' *Eurostat.* http://epp.eurostat.ec.europa.eu/portal/page/portal/income_social_inclusion_living_conditions/data/database.

Healy, J. D. 2003. 'Excess Winter Mortality in Europe: a Cross Country Analysis Identifying Key Risk Factors.' *Journal of Epidemiology and Community Health*: 784–789.

Swedish Energy Agency, 2012. 'Energy in Sweden 2012'. http://www.energimyndigheten.se/Global/Engelska/Facts%20and%20figures/Energy_in_sweden_2012.pdf.

November 2013

14. It is common practice in the UK to consider issues such as mould, condensation and damp as indicators of possible fuel poverty.

15. The average U-values are derived from the Building Performance Institute Europe's Data Hub for the Energy Performance of Buildings (BPIE 2013). This contains data on the average U-value of walls for single family dwellings built in different periods. This has been combined with data on the amount of floor-space in dwellings built in each period to calculate a weighted average U-value for each country's single family dwelling stock.

16. To reflect the fact that each country's climate is different (with colder climates necessitating lower U-values) we also considered the 'optimal' wall U-value for buildings in the UK and Sweden. Each optimum was calculated to reflect the most cost-efficient standard for buildings in each country to make their contribution to the EU's 2050 climate goals. Sweden is much closer to achieving its optimal U-value. See 'The Cold Man of Europe' report for more details.

Energy efficient buildings – beware possible health risks

An article from The Conversation.

By Melissa C. Lott – PhD Student: energy, environmental and public health trade-offs of energy system technology transitions, focusing on air pollution at University College London

The primary goal of home energy efficiency initiatives might be to reduce total energy consumption, but these projects could have a negative impact on public health if we do not take care.

Global climate change has been called the biggest global public health threat of the 21st century – and energy efficiency is a key tool in our efforts to reduce greenhouse gas emission levels.

Efficiency projects allow us to more effectively manage growing energy consumption without sacrificing services that we value. In the cost-optimised 2°C scenario set out by the International Energy Agency (the temperature rise that we have to stick within if we're to mitigate climate change), end-use efficiency improvements will be responsible for 38% of the global emissions reductions between now and 2050.

Without these emissions decreases, the World Health Organization expects 250,000 additional deaths to occur each year, caused by climate-related malnutrition, malaria, diarrhoea and heat stress around the globe.

Given these numbers, it seems logical to push forward with blanketing energy efficiency investments. However, there is evidence to show that we must take care in how we implement projects.

In a 2014 article published in the *British Medical Journal*, James Milner and his co-authors outlined how some home energy efficiency improvements could cost lives by increasing indoor radon exposure and the risk of developing lung cancer.

According to the authors, energy efficiency projects could lead to an estimated 56.6% increase in average indoor radon concentrations. They calculate that the corresponding increase in radon exposure could lead to 278 premature deaths (the equivalent of 4,700 life years lost) each year in the UK.

After smoking, radon exposure is the most important risk factor in developing lung cancer. This colourless gas, which occurs naturally from the indirect decay product of uranium or thorium, can be found in indoor air. It produces a radioactive dust that is trapped in our airways. This radiation then causes lung damage and increases the chance that we will develop lung cancer. Each year, an estimated 1,400 cases of lung cancer in the UK are primarily due to radon exposure, and about 21,000 in the US.

The increased radon concentrations in the Milner study stem from the fact that many energy efficiency improvements alter the way that buildings exchange indoor and outdoor air. These alterations are often aimed at reducing energy losses due to leaky windows or drafts around unsealed doors. In turn, these buildings can be more effectively heated and cooled, leading to observable public health improvements and decreases in total energy use.

However, they can increase some health risks. According to Milner and co-authors, while an individual project can be 'good for energy efficiency, indoor temperatures in winter and protection against outdoor pollutants, it has the potential to increase concentrations of pollutants arising from sources inside or underneath the home'.

A 2013 study suggested similar risks in retrofitted buildings from mould growth and 'sick building syndrome', where occupants appear to experience health issues from occupancy in a building. By trapping humidity inside the building, energy efficiency retrofits could unintentionally lead to dangerous mould growth. In turn, people in these buildings would be more prone to chronic fatigue, irritated lungs and watery eyes.

Using fans and other equipment to carefully control the indoor air quality could reduce or eliminate the negative co-impacts documented in these studies. Of course, the use of these technologies would offset some of the energy savings. But, they could also prevent an array of illnesses, which could stymie future energy efficiency proposals.

Energy efficiency projects can help to reduce total energy consumption. They are a key part of mitigating the impacts of global climate change. But we must be aware of any potential negative co-impacts on human health and take care to reduce their effect.

23 September 2014

⇨ The above information has been reprinted with kind permission from The Conversation Trust (UK). Please visit www.theconversation.com for further information.

The future of energy

UK future energy scenarios 2014: summary

The future of energy for the UK has never been so important.

Energy has become front page news; Electricity Market Reform, the General Election in 2015, environmental legislation, energy costs and developments in the economy will all have a major impact on the future energy landscape. No one can be certain how the energy future will evolve and this uncertainty may continue for decades.

Our Future Energy Scenarios (FES) represent transparent, holistic paths through that uncertain landscape to help the Government, our customers and other stakeholders make informed decisions.

The views of our stakeholders are essential in shaping the direction and content of our new future energy scenarios. Our goal is for everyone who has an interest in the UK's energy future to engage with us so that we can develop the most rich, robust and plausible range of scenarios possible.

Our future energy scenarios for 2014

This year we have broadened the range of our scenarios from two to four, which flex the axes of sustainability and affordability.

Power demand

Initial declines in peak demand are seen in SP and GG, driven mainly by energy efficiency.

Initial increases in demand in LCL are driven by increased consumer spending, economic growth and comparably less energy efficiency.

The adoption of electric vehicles and heat pumps will drive the changes in demand post-2025.

Growth in the later period is driven by increased numbers of houses in all scenarios.

Gas demand

By the mid-2020s, GG deviates from the other scenarios with fuel switching from gas in the residential and commercial sectors.

Demand in the commercial sector is variable, reflecting the potential for efficiency savings, new developments and power and gas prices.

All our scenarios reflect the underlying demand reductions across the industrial sectors.

There is an increasing requirement for gas-fired power generation to act as a backup for renewable generation in SP, GG and LCL.

Flexible power sources

The stronger regulatory and policy environment in GG and SP drives higher levels of interconnection. Weaker policy and regulatory approach in LCL and NP leads to lower levels of capacity.

Electricity storage has the potential to provide flexibility in balancing the system with increased levels of intermittent supply sources in the

Our future energy scenarios for 2014

Less money available ← **Affordability** → **More money available**

Low Carbon Life (LCL) is a world of high affordability and low sustainability. More money is available due to higher economic growth and society has more disposable income. There is short-term volatility regarding energy policy and no additional targets are introduced. Government policy is focused on the long term with consensus around decarbonisation, which is delivered through purchasing power and macro policy.

Gone Green (GG) is a world of high affordability and high sustainability. The economy is growing, with strong policy and regulation and new environmental targets, all of which are met on time. Sustainability is not restrained by financial limitations as more money is available at both an investment level for energy infrastructure and at a domestic level via disposable income.

No Progression (NP) is a world of low affordability and low sustainability. There is slow economic recovery in this scenario, meaning less money is available at both a government and consumer level. There is less emphasis on policy and regulation which remains the same as today, and no new targets are introduced. Financial pressures result in political volatility, and government policy that is focused on short-term affordability measures.

Slow Progression (SP) is a world of low affordability and high sustainability. Less money is available compared to Gone Green, but with similar strong focus on policy and regulation and new targets. Economic recovery is slower, resulting in some uncertainty, and financial constraints lead to difficult political decisions. Although there is political will and market intervention, slower economic recovery delays delivery against environmental targets.

Less emphasis ← **Sustainability** → More emphasis

future. However, significant cost reductions and access to multiple revenue streams are required to make storage a commercially viable solution.

Power supply

The GG generation background has adequate installed capacity to generate enough renewable electricity to meet the Government's renewable energy targets and sufficient low carbon generation to comply with the carbon budgets.

The focus on environmental targets within our SP scenario ensures continued support for the deployment of renewable generation but at a slower rate than Gone Green due to the less favourable economic conditions.

NP has adequate installed capacity to ensure security of supply with particular emphasis placed on cheaper forms of generation.

In LCL, the long-term decarbonisation strategy places increased emphasis on low carbon technologies, over renewables.

Gas supply

UK Continental Shelf production has a brief renaissance in the period up to 2020.

Norwegian gas continues to make up a significant part of the total supply to the market.

Import requirements vary significantly and uncertainty in the global gas market makes it difficult to predict whether the requirement for imported gas will be met by LNG or continental gas.

Our shale gas projections cover a wide range, reflecting government support and interest from producers, but also the currently unproven nature of the UK reserves.

Heat

Heat is a major proportion of energy use in our scenarios.

Savings from new A-rated boiler replacements are considerable and further savings are expected over the next decade.

New houses will be increasingly energy efficient as a result of incremental changes in building regulations.

There remains further efficiency savings from industrial and commercial sectors.

Heat pumps are assumed to initially be deployed in houses not connected to the gas grid, resulting in a net reduction in electricity demand.

Consumer

Lighting and appliances are the predominant consumers of electricity in the home.

The largest change in recent years and the near term in the residential electricity sector is around lighting and efficiency improvements instigated by European policy.

Increases in the number and size of some appliances will be offset by their efficiency improvements.

Consumer affordability combined with government incentives will determine the adoption of electric vehicles and heat pumps, which will drive the changes in demand in the future.

The impact of smart meters is small in comparison to the above changes and is bound by the speed of the roll-out and the adoption of time of use tariffs.

Transport

Range extended and plug-in hybrids are expected to dominate the electric vehicle market.

Users in London, second car and fleet buyers are likely to make up the bulk of the early adopters.

Market saturation of electric vehicles is not reached in any scenario.

Trial results showing the impact of time of use tariffs on electric vehicle

charging have been incorporated in our modelling. This has reduced potential peak demands.

2035–2050

The changing supply mix for electricity generation increasingly enables low carbon and renewable electricity to be used to meet electricity demand.

As well as meeting traditional demand, this electricity creates opportunity for heat and transport to be electrified in a sustainable way to realise environmental ambition.

Whilst heat can be electrified, it still needs gas for times of high demand, to constrain costs.

July 2014

⇨ The above information has been reprinted with kind permission from the National Grid. Please visit www.nationalgrid.com for further information.

Energy crisis looms for UK, France and Italy

New Anglia Ruskin report shows threat to EU countries from natural resource shortages.

A new report launched today [Friday, 16 May] warns that a number of countries in the European Union, including France, Italy and the United Kingdom, are facing critical shortages of natural resources.

Produced by the Global Sustainability Institute at Anglia Ruskin University, the natural resource maps indicate that some countries have less than a year of energy resources remaining and are almost entirely dependent on imports from the likes of Russia, Norway and Qatar.

By using the most recent data on known reserves and current consumption, the maps show that France has less than a year's worth of its own reserves of oil, gas and coal. Italy has less than a year of gas and coal, and only one year of oil. The UK fares only slightly better, with 5.2 years of oil, 4.5 years of coal and three years of gas remaining.

Some Eastern European members fare much better, with 73 years left of coal in Bulgaria and 34 years of coal in Poland. Meanwhile, Germany has over 250 years left of coal but less than a year of oil and only two years of gas.

By comparison, Russia has over 50 years of oil, over 100 years of gas and over 500 years of coal, based on their current levels of internal consumption.

Dr Aled Jones, Director of the Global Sustainability Institute at Anglia Ruskin, said:

'These maps show vulnerability in many parts of the EU and they paint a picture of heavily-indebted European economies coming under increasing threat from rising global energy prices.

'It is vital that those shaping Europe's future political agenda understand our existing economic fragility. The EU is becoming ever more reliant on our resource-rich neighbours such as Russia and Norway, and this trend will only continue unless decisive action is taken.'

Professor Victor Anderson of the Global Sustainability Institute added:

'Coal, oil and gas resources in Europe are running down and we need alternatives. The UK urgently needs to be part of a Europe-wide drive to expand renewable energy sources such as wave, wind, tidal and solar power.'

The full report, which also highlights issues such as food and water insecurity in North Africa and the Middle East, is available to download (http://www.anglia.ac.uk/ruskin/en/home/microsites/global_sustainability_institute/our_research/resource_management.html).

The maps are part of the Global Resource Observatory (GRO) project being carried out by the Global Sustainability Institute, which examines the relationship between the world economy and the environmental factors and resources it depends on. The GRO project has been generously supported by the Peter Dawe Charitable Trust.

The full GRO database of social, environmental and economic data will be made freely available to download this summer.

16 May 2014

⇨ The above information has been reprinted with kind permission from Anglia Ruskin University. Please visit www.anglia.ac.uk for further information.

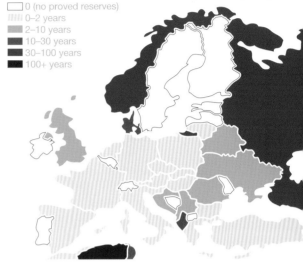

European oil reserves

Years left (proved reserves divided by consumption)
- 0 (no proved reserves)
- 0–2 years
- 2–10 years
- 10–30 years
- 30–100 years
- 100+ years

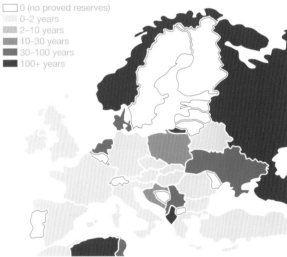

European natural gas reserves

Years left (proved reserves divided by consumption)
- 0 (no proved reserves)
- 0–2 years
- 2–10 years
- 10–30 years
- 30–100 years
- 100+ years

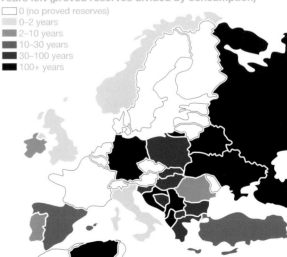

European coal reserves

Years left (proved reserves divided by consumption)
- 0 (no proved reserves)
- 0–2 years
- 2–10 years
- 10–30 years
- 30–100 years
- 100+ years

Smart thinking

Government figures suggest that, through to 2030, the net benefit of the smart meter programme to the UK will total £6.2 billion.

By 2020, some 30 million UK homes and small businesses will have smart meters installed under government plans to give consumers greater control over their energy use. Phil Sheppard, National Grid's Head of Network Strategy, explains the benefits of smart metering and how the wider creation of smart energy systems to link energy sources and users in a more intelligent way can help to deliver a sustainable future.

Technology has a vital part to play in solving the future energy challenges that we face as a nation in all sorts of areas, from security of supply to sustainability and affordability.

If we want to use our energy resources in the smartest way possible – and in the best interests of consumers – we need to harness new technology in order to get there more quickly and more efficiently. The advent of the smart meter in homes and offices is one example of this new technology, but in reality the roll out of the smart meter is just the start of our journey.

Smart meter roll-out

The idea of installing smart meters across the UK has been around for some time. The Government launched a three-stage process to upgrade the energy system using smart meter technology:

⇨ Stage one involved policy design and this was completed in 2011

⇨ A so-called Foundation stage is under way, whereby systems are being built, tested and installed, followed by...

⇨ The main installation programme, which is due to begin towards the end of 2015.

Installed by energy suppliers under a centrally controlled programme, smart meters will bring a number of benefits.

Consumers will be able to monitor their energy use in real-time and potentially make savings by connecting to new 'time-of-use' tariffs and using more energy when demand is lower, such as running dishwashers and washing machines overnight. Last but not least, we will see the end of estimated billing.

There are other factors to consider beyond consumer behaviour. For example, smart meters will allow us to understand the impact at a domestic scale of things like wind turbines, heat pumps and solar photovoltaics (PV). So, in effect, we will have a much more rounded picture of how we produce and consume energy.

Analysing the potential benefit to the UK of the smart meter programme is complex, but government figures suggest that, through to 2030, the net benefit to the UK will total £6.2 billion, even once the estimated £10.9 billion programme cost is accounted for.

Beyond smart meters

The implications stretch beyond homes and businesses, however. Building new infrastructure such as power stations is expensive and time-consuming, so finding new ways to reduce and make better use of demand, especially at peak times, makes economic as well as environmental sense.

From a transmission perspective, smart meters bring benefits for customers by giving us new ways to optimise use of demand and micro generation, as well as more cost-effective choices for balancing the grid and reducing the need for network investment.

Of course, smart meters in isolation cannot hope to reshape our collective energy consumption; they are just part of the equation. At National Grid we have a stake in the smart meter programme because we are right at the centre of the UK's whole energy system – joining everything together. The concept of smarter thinking that underpins the programme is mirrored in the way that our transmission network operates.

We are working with industry to deliver the right solutions at the right time, developing and deploying rules, tools, technical and commercial solutions to minimise the amount of network investment that is required, while pushing the system further in operational timescales and increasing automation.

This is enabling us to progress work on new assets such as the Western Link, a £1 billion project being developed by National Grid and Scottish Power to bring renewable energy from Scotland to England and Wales via the construction of a high-voltage direct current cable.

The role of smart energy systems

If we want to optimise the use of energy resources in the best interests of consumers and manage the transition to 2050, we need to think beyond the grid itself and the information and communication technology infrastructure that underpins it and start to consider smart energy systems.

This begs the question 'what do we mean by smart energy systems'? There are many definitions, but to my mind what we are really talking about is finding ways to make the way we generate and consume energy more interconnected.

Renewable forms of energy are less flexible than fossil fuels, which allow us to store large amounts of energy, so we must find economic ways of capturing and using excess energy from the sun, tides and wind.

A smart energy system is all about using this renewable energy in an efficient and affordable way, as well as getting the best end result for consumers through developments such as smart electricity grids, thermal storage and smarter ways to connect the electricity, heating and transport sectors.

The only way that these smart energy systems can be developed effectively is through collaboration and co-operation right across the sector: government, regulators, network operators, distribution companies, consumer groups and of course National Grid – we all have a contribution to make.

The smart meter roll out lays the foundation; now it's up to all of us to build on this and make sure that consumers get the best deal for the energy they use, through systems that really work for them.

Read more:

⇨ Stephen Marland, National Grid Gas Demand Manager on the implications of alternative technologies to our homes and the energy sector: http://www.nationalgridconnecting.com/heating-our-homes/.

⇨ 'The future of heating: meeting the challenge' was published last year and sets out specific actions to help deliver low carbon heating across the UK in the decades to come: https://www.gov.uk/government/uploads/system/uploads/attachment_data/file/190149/16_04-DECC-The_Future_of_Heating_Accessible-10.pdf.

20 June 2014

⇨ The above information has been reprinted with kind permission from the National Grid. Please visit www.nationalgridconnecting.com for further information.

What are smart meters?

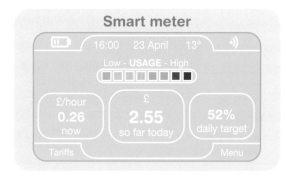

Smart meter

16:00   23 April   13°

Low - **USAGE** - High

£/hour
0.26
now

£
2.55
so far today

52%
daily target

Tariffs   Menu

Energy companies have started installing smart meters in homes across Great Britain – most homes should have one by 2020. Smart meters are the next generation of gas and electricity meters and they can offer a range of intelligent functions.

For example they can tell you how much energy you are using through a display in your home. They can also communicate directly with your energy supplier meaning that no one will need to come and read your meter in future.

Most of the smart meters that are being installed today use mobile phone-type signals to send meter readings to your supplier, and other wireless technologies to send information to the in-home display.

The Government is requiring energy companies to install smart meters for their customers, and is setting out rules to ensure that they do this in a way that is in the interests of consumers, including rules around data access and privacy, security, technical standards for the smart metering equipment and meeting the needs of vulnerable consumers.

How smart meters work

Smart meter monitors your everyday energy usage.

Readings are automatically sent wirelessly to your energy provider.

Get sent an accurate energy bill without having to submit any meter readings.

You can even view a detailed energy report showing your usage online.

⊕ Pros

- Receive accurate energy bills without having to submit a meter reading (you won't get estimated bills).
- Shows exactly what you are using, which in turn will help you make choices where to save.
- More in control of your energy costs.
- Detailed insight into your energy consumption.
- It'll be easier to switch energy providers.

⊖ Cons

- Potential problems ensuring the security of metering data.
- Protecting the privacy of personal information.
- Challenge of verifying that the new meter is accurate.
- Cost of new meter.
- Potential problems transitioning to new technology and processes.
- Concern of possible health effects of wireless technology.

Sources: Smart meters, *GOV.UK*, What are smart meters?, *British Gas 2015*, Smart Meters Also Pose Challenges, *About.com 2015.*

Fewer than half of consumers want a smart meter

New research suggests households are not fully behind the Government's £11 billion nationwide roll-out of smart meters.

By Michele Martinelli

According to research carried out by the Smart Meter Central Delivery Body (SMCDB), less than half of consumers want to have a smart meter installed in their homes.

SMCDB, an organisation created to promote smart meters, found that 84% of people had heard of smart meters, however, just 44% said they would be interested in having one installed.

To date, approximately one million meters have been installed, primarily by British Gas, with the full roll-out to start in late 2015.

UK roll-out by 2020

The Government-led smart meter roll-out is projected to cost £11 billion and is expected to be completed by 2020. Energy companies are responsible for installing smart meters; however, customers can refuse to have one installed in their home.

Advocates of smart meters say the devices will help customers gain control of their energy bills, as well as put an end to estimated energy bills. Smart meters automatically send readings to energy suppliers and allow users to better understand their energy consumption, by providing them with real time usage figures.

Most consumers don't trust energy suppliers

The research also found that more than half of energy customers did not trust energy suppliers. Chief Executive of the Smart Meter Central Delivery Body, Sacha Deshmukh said that smart meters would be vital in restoring trust to the energy market.

'Antiquated systems for recording energy use and managing billing are no longer fit for purpose. Households need to be able to take control of their energy use and bills,' said Deshmukh.

[Smart meters] will create newly empowered consumers

Speaking on the research, Deshmukh said: 'Households need to be able to take control of their energy use and bills. For this to happen, the national smart meter roll-out is the essential transformation of the technology we use to buy energy.

'It will create newly empowered consumers, and increase trust in those who sell us gas and electricity – and our research bears this out.

'Almost half of consumers told us that they are interested in having a smart meter installed in their homes. That is why it is so important that government is driving forward the programme to install smart meters across Great Britain and has brought together all the electricity and gas suppliers and networks to deliver this critical upgrade to the energy infrastructure in all of our homes.'

'More in control and better able to manage their energy'

Ann Robinson, Consumer Policy Director at uSwitch, said: 'With trust in the energy industry at an all-time low and people finding it difficult to afford the energy they need, it is high time that we now get on and deliver the smart meter programme.

'Our own research backs up today's findings – that people using smart meters have a higher level of trust in their energy suppliers and experience greater customer satisfaction. They also feel more in control and better able to manage their energy usage.

'Energy companies are amongst the worst offenders for getting bills wrong – in fact a fifth of homes have received incorrect bills over the last two years. With energy accounting for the biggest chunk of household spend after the rent or mortgage, this is incredibly frustrating. The move to smart meters will ensure that people receive accurate, up-to-date bills based on their actual usage. But in the meantime it's important that we all continue to provide our suppliers with regular meter readings.'

9 June 2014

⇨ The above information has been reprinted with kind permission from uSwitch. Please visit www.uswitch.com for further information.

Fracking threat to the UK

Fracking is a nightmare! Toxic and radioactive water contamination. Severe air pollution. Tens of thousands of wells, pipelines and compressor stations devastating our countryside and blighting communities. All while accelerating climate change. And to produce expensive gas that will soon run out. So why do they want to do it?

Extreme energy

Fracking is just a symptom of a much wider problem. As easier to extract energy resources are exhausted by the unsustainable energy consumption of the present system, we are resorting to ever more extreme methods of energy extraction. Over the last century the exploitation of fossil fuels has moved from tunnel mining for coal and drilling shallow oil wells to tearing apart whole mountains and drilling in a mile or more deep of ocean.

As existing energy resources deplete the default response has just been to try harder; dig or drill deeper; go after lower quality resources or move on to more remote locations. This increasing effort has consequences: increasing pollution, more dangerous working conditions, greater greenhouse gas emissions, more land use and less resources available to other sectors of society.

At present we are on a course which leads towards a world dominated by

energy extraction, one where most of the energy produced is used to run the extraction processes while people live and die in its toxic shadow. The present system's addiction to massive amounts of energy is driving this headlong rush towards oblivion and unless something is done to stop it we will all be dragged down into hell with it.

Tar Sands, Mountain Top Removal, Deep Water Drilling, Biofuels and Fracking are all symptoms of this scramble to suck the last and most difficult to reach drops out of our planet. Even more extreme extraction methods are being contemplated such as Oil Shale and Methane Hydrates, while existing methods are slowly growing more extreme as easier to extract resources are depleted.

Unconventional gas

The UK unconventional gas (and to a lesser extent oil) extraction is the main new threat, in the form of three different processes; Shale Gas, Coal Bed Methane (CBM) and Underground Coal Gasification (UCG). While there are a lot of differing technical details these processes all involve drilling large numbers of directional wells at regular intervals, coating the landscape.

The scale of these new more intense methods are like nothing we have seen before. Up until now the largest onshore gas field in the UK, Saltfleetby in Lincolnshire, had only eight wells. To produce the same amount of unconventional gas would require hundreds of wells to be drilled. To temporarily replace just one offshore North Sea gas field would require thousands of unconventional wells.

As well as requiring many more wells these methods also involve much more. Shale Gas and Oil require massive, slickwater hydraulic fracturing, to be carried out on every well. Millions of gallons of water, sand and chemicals are injected under massive pressure. CBM wells are also often fracked. UCG which involves setting fire to coal seams underground is even more extreme.

These unconventional wells also have much shorter lifespans, with production from a typical shale well declining by 70 to 80 per cent in the first year alone. This means that large numbers of new wells need to be constantly drilled to maintain production, even for short periods. In many areas of the US, unconventional gas is already peaking after less than a decade of exploitation.

Impacts of unconventional gas (and oil)

For all these processes water contamination is a major issue. All wells will eventually leak, as steel casings rust and cement rots, and unconventional gas (and oil) means many, many more wells. Contamination of groundwater has been a consistent feature of unconventional gas extraction, in the US, Canada and Australia.

The amount of water used in these processes and the amount of waste produced are also major issues. In Colorado, farmers are losing access to water as fracking companies buy up supplies. Meanwhile the vast streams of toxic and radioactive waste are a nightmare to dispose of, and attempts to get rid of this waste by injecting it into the ground are causing large numbers of earthquakes.

Air pollution is also an underappreciated threat from unconventional gas. In previously pristine wilderness areas of the US, ozone levels now routinely exceed those in the centre of Los Angeles, while leaking toxic and carcinogenic hydrocarbon vapours are also common. Such pollution can be blown hundreds of miles from its source. Breathing difficulties are common complaints for those living in the shadow of these industries.

While targeted health studies of the effects of these developments have just not been done, what evidence there is shows major impacts. Cancer clusters, neurological and reproductive problems in humans and animals have all been reported and should be expected given the chemicals that are being emitted. In the vicinity of unconventional gas extraction, communities are getting sick and the response has been make people prove that the industry is the cause, or shut up.

Climate catastrophe

At a global level, there are already far more conventional fossil fuel reserves that we can afford to burn without causing catastrophic climate change. As with all unconventional fossil fuels unconventional gas (and oil) simply adds to this store of unburnable carbon. Widespread exploitation of unconventional fossil fuels could produce enough carbon dioxide to make the planet literally uninhabitable.

In the shorter term, methane emissions from these processes amplify the effects of the carbon dioxide emitted. Studies have shown that Shale Gas and CBM are worse than burning coal in the short term, and it is the short term that matters when considering potential tipping points in the climate system like melting arctic permafrost and the fate of the Amazon rainforest. UCG is even worse, with its direct carbon emission far higher than from the conventional exploitation of coal.

What can be done?

While all this may seem very bleak, there are rays of hope within this dark cloud. Unconventional fossil fuels are much more dispersed than conventional ones, meaning that in order to get them, many more communities are affected but must at least passively consent to their extraction. If these communities get organised to resist this invasion then it can be stopped. This is already happening is many places across the globe (for instance in Australia) but everyone needs to do their bit if this juggernaut is to be stopped.

Want to get organised? Want to take action? Get stuck in… (http://frack-off. org.uk/get-stuck-in/).

⇨ The above information has been reprinted with kind permission from Frack Off. Please visit www.frack-off.org.uk for further information.

Poll shows 63% not opposed to fracking

The British public remains deeply divided over whether fracking should be allowed to go ahead in the UK despite the Government's attempts to promote it as a way of improving energy security and cutting prices, new research suggests.

A survey by Opinium for Govtoday has revealed that 37% of adults are opposed to the exploitation of shale gas reserves while only 25% are in favour. However, as many as 28% describe themselves as neither in favour nor opposed to the controversial drilling technique, and another 10% 'don't know'.

The results, published on the day of Govtoday's Fracking conference in London, show that scepticism among the public is growing – with a previous poll in August 2013 putting support for fracking higher at 32% and opposition at the lower figure of 33%.

The polling also shows the effect of nimbyism, with support for fracking falling to just 19% when respondents were asked whether they would be happy to see it in their local area. In that scenario, opposition rises to 45%. Even so, more than a quarter, 26%, continue to be neither in favour nor opposed.

Those who support fracking most often cite reducing the UK's dependency on foreign energy sources, and the creation of jobs and economic growth, as the main reasons for their position, the research shows. Among opponents, damage to the local environment, as well as noise and disruption for residents, are the main concerns.

David Cameron has previously said his government would go 'all out for shale' and offered councils the opportunity to keep 100% of the business rates from fracking sites in a bid to shore up support. A portion of the profits from drilling is also set to go straight back into local communities.

The Prime Minister has claimed fracking could create 74,000 jobs and bring in £3 billion of investment. But the attempt to win over the public does not appear to be working.

According to the survey, 44% of people admit to knowing only 'a little' about fracking, with 17% knowing nothing about it and 9% having not heard of it at all. Just 24% of Brits say they know a 'reasonable amount' about the process and 6% say they are well informed.

Support for fracking is highest on the right of politics – with 44% of Conservative voters and 38% of UKIP supporters in favour. Shale exploitation has the backing of only 19% of Labour voters and 24% of Liberal Democrats.

Meanwhile, asked what would change their mind about fracking, 47% of those currently opposed cited proof that the drilling would not cause harm to groundwater and the environment.

Some 39% said they might change their view if it was shown fracking did not cause any earthquakes, and 32% said they would alter their view if it meant lower energy bills. More than a quarter, 28%, said they would be more likely to support fracking if nearby residents received a share of the profits.

16 May 2014

⇨ The above information has been reprinted with kind permission from Govtoday. Please visit www.govtoday.co.uk for further information.

Facts and information for nuclear energy

Nuclear power is a proven form of electricity generation worldwide.

Nuclear supplies more than 11% of the world's electricity. Today, the world produces as much electricity from nuclear energy as it did from all sources combined in 1960.

There are currently more than 430 commercial nuclear reactors operating in 48 countries, with a further 173 under construction.

Nuclear energy currently supplies around 19% of the UK's electricity

Nuclear energy has supplied up to a third of the UK's electricity since 1956. Today 18 reactors on nine sites supply nearly a fifth of the country's electricity.

UK public opinion

Public support for nuclear energy, alongside other low-carbon sources, has been strong for several years.

Nuclear comes first in the energy sources the public believe is needed to keep the lights on, but industry must not take this support for granted. The case for an expansion of low-carbon nuclear energy remains compelling, public confidence in the safe operation of nuclear plants and the safety standards of the industry, must be maintained.

World class, competitive industry of highly skilled professionals

The civil nuclear industry employs more than 54,000 highly skilled people throughout the UK, with more than 80,000 jobs being directly or indirectly linked with the industry. Companies in the UK nuclear industry have the capability to provide more than 80% of the work involved in new nuclear power station projects in the UK. This represents billions of pounds of investment into the UK economy.

Over the next two decades it is forecast that globally there will be £930 billion investment in building new reactors and £250 billion in decommissioning those coming offline. The nuclear new build programme in the UK alone could generate up to 40,000 jobs in the sector at its peak. The nuclear industrial strategy sets out the basis for a long-term partnership between government and industry to exploit those opportunities.

The industry has the experience to operate and maintain them throughout their full lifecycle.

There are hundreds of successful businesses in the UK civil nuclear industry. Find local businesses in your area on the NIA Jobs Map (http://www.niauk.org/nia-industry-maps).

The world has changed dramatically over the last few years

Security of supply concerns have grown and energy prices across the board have risen sharply, pushed up by global oil and gas prices. Climate change still remains high on the political agenda with a number of targets due to be met over the coming years. In the UK, CO_2 emissions have started to fall but remain high by international standards. Progress on energy efficiency and renewables uptake has been slow and the looming energy gap draws closer with North Sea oil and gas resources running out faster than expected.

A balanced energy mix will help to ensure security of energy supply in the UK

Nuclear energy provides Britain with energy security by reducing the country's dependence on foreign imports. Nuclear provides 'baseload' electricity, the minimum amount of energy needed to power the grid, without air pollution and exceptionally low-carbon emissions (similar to wind).

All too often the green energy debate is framed as 'nuclear vs renewables'. In fact there is strength in diversity. A mix of low carbon technologies, together with Carbon Capture and Storage, and alongside efficiency measures, are needed to enable the transition from fossil fuels while meeting climate and energy security goals.

Nuclear fuel can provide more energy

An inch-long pellet of Uranium-235 has more energy than one tonne of coal. A fission reaction is about a million times more energetic than a chemical reaction such as the burning of coal, oil or gas. One gram of uranium-235 when undergoing fission in a nuclear reactor gives energy roughly equivalent to burning three tonnes of coal or two tonnes of oil.

Nuclear power is efficient

In comparison to a traditional fossil fuel such as coal, nuclear fuel generates millions of times more power. For example 1kg of coal would power a 60-watt light bulb for four days. In comparison, nuclear would power the same bulb for 685 years.

Nuclear also saves billions of tonnes of CO_2 emissions. Compared to coal, the power generated by existing nuclear power stations avoids 40 million tonnes of CO_2 per year – the equivalent to taking half of Britain's cars of the roads.

The health effects of such a dramatic reduction in fossil fuels are measurable. A recent paper reported in _WNN_ (Life-saving case for nuclear, 3

April 2013, http://www.world-nuclear-news.org/EE_Life_saving_case_for_nuclear_0304131.html) said 1.8 million lives have been saved through the use of nuclear power instead of a representative mix of coal and gas.

Nuclear is a low-carbon source of energy, making it a valuable contributor in the battle against climate change

In the UK, the 2008 Climate Change Act set a legal obligation to reduce greenhouse gas emissions by 80% from 1990 levels by 2050. To help meet this ambitious target, the Government has embarked on a policy of Electricity Market Reform (EMR) to stimulate investment in low-carbon energy by private sector developers.

CO_2 is the main cause of climate change. Nuclear power generation is extremely low carbon. Through its life cycle, nuclear power produces a similar amount of CO_2 emissions as wind power, and much less than solar. For each kilowatt hour of nuclear electricity, five grams of carbon dioxide is emitted, compared to around 365g from gas-fired power stations, or 900g from a coal-fired power station.

In the UK, the power generated by existing power stations avoids the emissions of 40 million tonnes of carbon dioxide a year – the equivalent of taking almost half of Britain's cars off the road.

Nuclear energy is an economically competitive form of power generation

The price of fuel represents a small fraction of the total operating costs for nuclear power. As a result, these costs are stable and predictable, unlike fossil fuels where prices can fluctuate.

For both nuclear and renewables, the economics are dominated by the construction costs. Government analysis shows nuclear to be the cheapest form of low-carbon electricity generation.

The UK needs substantial investment in generating capacity over the next 20 years

Maintaining Britain's energy security is an urgent and immediate priority. Many of the current nuclear power stations will close by 2025, while demand for electricity is expected to rise, and as Britain's North Sea oil and gas resources decline. As low-carbon technologies have long lead times, investment in new generation is needed now, if it is to come on stream in the early 2020s.

There are major socio-economic gains to be made from Britain's energy challenge: in terms of rebalancing the economy towards manufacturing and construction, boosting supply chains, and creating large numbers of jobs. There are challenges to achieving a new build programme, but government, regulators and industry are working together to overcome these. The workforce and supply chain companies, in the UK and world-wide, are capable of delivering a fleet of new reactors.

With 58 operating nuclear reactors, France is the world's largest electricity exporter. France has proved a large scale energy transition can be achieved. In response to the 1970s oil crisis, France built an entire nuclear fleet in a decade, and today has a carbon intensity of around $70gCO_2$/kwh. The UK currently has a carbon intensity of around $450gCO_2$/kwh, due to its reliance on fossil fuels.

Strengthening the UK industrial base and creating thousands of new jobs

During the construction of new build nuclear power generation thousands of jobs will be created.

New nuclear build in the UK is achievable

Support for nuclear energy has increased with more than 70% of the British public believing nuclear power should form part of a balanced energy mix for the future (YouGov, November 2012). There are challenges to achieving a new build programme, but if government, regulators and industry work together, these can be overcome. The workforce and supply chain companies, in the UK and world-wide, are capable of delivering a fleet of new reactors.

Modern reactor designs generate electricity 90% of the time, even when the Sun does not shine or the wind does not blow

New reactor designs are being licensed and built in other countries, such as Finland, France, China and India and South Korea. New nuclear plants are expected to generate around 90% of the time (compared to 35% for wind power, or 10% for solar in northern Europe). Nuclear is

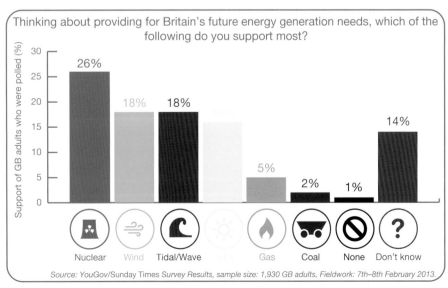

Thinking about providing for Britain's future energy generation needs, which of the following do you support most?

Source: YouGov/Sunday Times Survey Results, sample size: 1,930 GB adults, Fieldwork: 7th–8th February 2013.

therefore an essential element in the mix to provide clean, reliable baseload energy, and offers a competitive return alongside other low-carbon generation.

New reactor designs, such as small modular reactors are smaller, quicker and should prove cheaper to build once a major programme is established. They use the latest technology and safety standards, and use the nuclear material more efficiently to generate more power and less waste than older designs.

It is worth noting that the costs for emerging technologies such as marine and CCS do not make them economic at the moment, and are as yet unproven on a commercial scale.

Modern nuclear power stations are amongst the most robust, secure structures in the world

Power stations are extremely robust structures and have a multi-layered defence against possible terrorist attacks. The Office for Nuclear Regulation is responsible for approving security arrangements within the industry and enforcing compliance.

Even after the Tohoku earthquake and tsunami in March 2011, most of the nuclear power stations in the affected region were shut down safely, though the four oldest reactors did suffer considerable releases of radioactive material.

However, we should emphasise the differences between Japan and the UK. The UK is not near any major seismic fault-lines and is not susceptible to the magnitude of earthquakes experienced in Japan. UK plants are protected against the effects of a one in 10,000 year's earthquakes meaning that even if they were hit by the worst earthquake that could be expected in 10,000 years, our plants would be safe.

Nuclear fuel supplies are assured

Accessible and affordable uranium ore from known reserves in countries like Canada and Australia can be assured for the full lifetime of a fleet of new UK reactors. The required amount of fuel is small in volume and is easily stored, indeed it would be technically possible to store all the fuel for the lifetime of a nuclear reactor on site if necessary. A rapid expansion of nuclear power on a worldwide scale would not overturn this position.

The nuclear fuel cycle

There are various stages of the nuclear fuel cycle: from initial mining, through fabrication, electricity generation, reprocessing and eventual final waste disposal.

The UK has managed its radioactive waste safely for over half a century

All of the UK's nuclear waste is securely contained with an ever-increasing amount being solidified to make it suitable for long-term management. Other countries have already demonstrated how safe and secure long-term management and permanent disposal of nuclear waste is feasible. Countries such as Finland and Sweden are in the process of constructing deep geological radioactive waste disposal sites, following successful government and public consultation.

The UK Government has also set out plans to use deep Geological Disposal Facilities for high level waste following firm recommendations from a Committee on Radioactive Waste Management (CoRWM) report in 2006. The Nuclear Decommissioning Authority has taken the lead in establishing a site for the repository and is in the process of finding a suitable location.

If a fleet of new plants were commissioned to replace the current ones, they would only add around 10% to the volume of existing waste over their 60-year lifespan.

Where does radiation come from?

Our average annual radiation dose in the UK is 2.6 mSv. Typically, we get about 85% of it from natural sources. Other sources include cigarette smoke, and air pollution from fossil fuels used in transport or coal-fired power stations, for example.

How can we use radiation?

Radiation has changed our lives, and led to dramatic advances in medicine, agriculture, industry, energy production and research.

The UK's civil nuclear programme has a world class, strong, independent regulator

All facilities are licensed by the industry regulator the Office for Nuclear Regulation (ONR), an independent organisation, formally part of the Health and Safety Executive. Modern reactors have reliable advanced safety systems – and any imbalance in the normal system operation will lead to automatic shutdown.

A nuclear timeline for the UK

The UK was an early pioneer in nuclear power generation. A wide range of experimental and ground-breaking reactors were developed, which have helped to push the bounds of technology. Key events in the nuclear timeline include the discovery of Uranium in 1789 as well as the recent strike price agreement of 2013.

⇨ The above information has been reprinted with kind permission from the Nuclear Industry Association. Please visit www.niauk.org for further information.

Cumbria's nuclear dump can't bury the waste problem

The region's complex geology has already been rejected once. More astonishingly, 60 years of the nuclear age has yet to produce a single deep disposal site anywhere on Earth.

By Damian Carrington

Nuclear waste is long-lived and extremely dangerous and consequently poses near interminable and politically toxic questions. Wednesday's vote, by councils in Cumbria on whether to move forward with plans for deep underground storage of waste in the area, is the perfect illustration.

But let's start at the beginning. Is burying nuclear waste deep within bedrock necessary? Yes is the answer of most experts, though only because it is the least worst option. Over 25 countries with significant waste piles have opted for deep disposal as the ultimate solution. The perils of floods, terrorist attacks and earthquakes mean long-term surface storage poses even greater risks than entombing the waste in a rock sarcophagus. As for blasting it into space, as some have suggested, does placing tonnes of radioactive waste on top of hundreds of tonnes of high explosive sound smart?

So is deep disposal safe? The answer is no-one knows, because despite 60 years having passed since the nuclear age began, no-one has ever done it. Sweden has the most advanced plans, with an experimental deep disposal facility, but no nuclear waste has yet been consigned to its depths. In the US, vast amounts of money was spent over many years but ended up with plans for deep disposal at Yukka Mountain in Nevada being abandoned in 2010.

The situation is uncomfortably similar in the UK. In the 1980s a nationwide search for a suitable burial site was undertaken. In a process described by Professor Stuart Haszeldine, a geologist at the University of Edinburgh, as 'mysterious and which no-one understood', the site selected was conveniently very close to the epicentre of the UK nuclear industry at Sellafield in Cumbria.

The Government then spent £400 million on 22 boreholes investigating the Longlands Farm site but a 65-day public planning enquiry ultimately rejected it because the site was technically unsuitable, as well as the proposed project having severe impacts on the surface.

They key problem that has to be solved with deep disposal is ensuring, for millennia, that radioactive waste does not leak out into groundwater which is then brought to the surface. The UK is fortunate not to suffer serious earthquakes but the site chosen had extremely complex geology, riven with fractures, meaning there was an unacceptable risk of radioactive waste being washed back to the surface.

Extraordinarily, after all that time and money was spent in ruling out that site, it is back in the running under the plans the councils are voting on. Professor Haszeldine is scathing about how this has happened: 'This is a very short-sighted policy, run by driving local councils into volunteering for the wrong reasons: financial inducements. Many of the statements being made by the authorities [about the site] are misleading to wrong.'

In fact the situation now is even worse than in the 1980s. The proposal now would see spent fuel and high-level waste buried, rather than only intermediate-level waste. The former is much hotter than the latter, which would hugely accelerate the circulation of groundwater, according to Professor Haszeldine.

'The waste exists and deep geological disposal is by far the best solution,' said Professor Haszeldine. 'But we should have a genuine national search for technically qualified sites. Ultimately, do we believe in evidence-based policy or political opportunism to exploit communities with limited economic opportunities?'

A number of worrying conclusions can be drawn from all this. First, the process of choosing a deep disposal site appears more concerned with burying political problems than finding the right geological site. Asking local communities for their assent is of course essential, but getting that assent by burying geological problems under layers of cash will only see those problems rise, zombie-like, from the grave in future.

Second, after half a century of nuclear power and its ever-growing waste pile, isn't it astonishing that no-one anywhere on the planet has demonstrated a permanent way to deal with it? All the while, the costs to the public of keeping this toxic legacy safe on the surface continue to spiral out of control. Yet, despite only having sticking plaster solutions to the waste problem to hand, the UK Government and others are dead set on building a new fleet of reactors.

Perhaps most worrying of all is that whatever the result of the votes in Cumbria, ministers will undoubtedly keep digging themselves deeper into the nuclear hole. Dealing with legacy waste is unavoidable but adding to this apparently unsolvable problem is not. Phasing out nuclear power is challenging in terms of energy policy, but a breeze compared to the long-term waste problem. The backing of new reactors once again shows the triumph of the short-term political fix over the genuine long-term solution.

30 January 2013

Burning land: how much land will be required for Europe's bioenergy?

New research reveals the vast land footprint for Europe's bioenergy needs.

By Ariadna Rodrigo and Robbie Blake

New research by the Vienna University of Economics and Business (WU) for Friends of the Earth Europe shows how Europe's consumption of bioenergy is exerting unparalleled and unsustainable new pressures on the world's land and forests.

With demand for bioenergy (crops and wood used for transport fuel, heating and electricity) set to more than double between now and 2030, Europe will require an area of land and forest the size of Sweden and Poland combined to meet our bioenergy needs.

Our consumption of land is out of control. Europe is already the continent with the highest dependence on land from outside its borders, and has the second highest level of land consumption per capita. Our growing hunger for unsustainable bioenergy, encouraged by subsidies, targets and policies, will only exacerbate the disproportionate pressure Europe exerts on precious global land resources.

It is essential that the EU starts to take account of its land footprint by measuring it and setting reduction targets – alongside carbon, water and materials. The European Commission already committed to start measuring and reducing Europe's natural resource consumption in its 2011 *Roadmap to a Resource Efficient Europe*.[1]

In its energy policy, the EU must urgently cap and set a trajectory to phase out bioenergy that competes with food (including crop-based biofuels). The 2030 climate and energy package must limit the use of bioenergy to sustainable levels, and support sustainable bioenergy, e.g. from waste residues, ensuring

1. http://ec.europa.eu/environment/resource_efficiency/pdf/com2011_571.pdf

that they deliver greenhouse gas emissions reductions – within ambitious climate change, renewable energy and energy savings targets.

The scale of Europe's land consumption from bioenergy

According to the WU calculations of Europe's appetite for land for bioenergy:

⇨ The land footprint of bioenergy consumption in 2010 was 45 million hectares, equivalent to the entire land area of Sweden. By 2030, if current trends continue, EU bioenergy consumption is expected to increase by 58%, occupying 70 million hectares of land – equivalent to the size of Sweden and Poland combined.

⇨ In 2010, bioenergy accounted for 8% of all energy consumption; this is set to increase to 12% by 2020 and even more by 2030. However, there are important negative social and environmental consequences attached to some forms of bioenergy on a large scale, particularly land competition and carbon emissions.

⇨ The use of wood to generate heating and electricity comprises the biggest land footprint – roughly 30 million hectares of forest in 2010, and is set to expand to 40 million hectares by 2030. If all of this wood came from inside the EU, it would require nearly 40% of all Europe's productive forest area.

⇨ Unless the EU successfully reforms its biofuels policy, land use requirements for biofuels

will increase substantially. Between 2010 and 2020 there would be an increase of 130%, meaning 11 million hectares of agricultural land will be dedicated to biofuel production – an area the size of Bulgaria or the whole of Germany's agricultural land area.

What's driving Europe's bioenergy land demand?

Current EU subsidies, targets and policies have led to this increase in use of bioenergy, and unless policy change can limit bioenergy use to sustainable levels and production, the EU bioenergy land footprint will continue to increase dramatically.

Friends of the Earth has documented[2] how increasing demand for limited land to produce bioenergy is already causing land grabs, conflicts, deforestation, biodiversity loss, climate change and high food price volatility, which threatens global food security. Expansion and intensification of unsustainable agriculture and forestry practices for bioenergy will contribute to the further destruction of vital ecosystems, and can further increase carbon emissions.

A growing land footprint for bioenergy will be met at the expense of the resources of other nations and the quality of life of their citizens, as well as of forests, grasslands and other natural habitats. It makes Europe dependent on the availability of cheap and abundant land around the world for our energy security.[3]

2. FoEE (2013) Commodity Crimes http://www.foeeurope.org/commodity-crimes-211113; FoEE (2010) Africa Up For Grabs http://www.foeeurope.org/agrofuels/FoEE_Africa_up_for_grabs_2010.pdf

3. Mostly coming from USA, Canada, Russia, Brazil for wood; Argentina, Brazil, USA and southeast Asia for biofuels.

The EU framework for climate and energy towards 2030 will therefore be crucial in determining bioenergy demand and supply, which will have considerable implications for land pressure and resource use both inside and outside Europe. The future framework must include robust mechanisms to limit the use of biomass for energy to sustainable levels and sustainable land use and forest management; ensure efficient use of biomass and land resources in line with the principle of cascading use;[4] and guarantee genuine GHG savings.

Europe's overconsumption of land

Whilst bioenergy is a significant driver of land use expansion, it's not the only one (another is meat consumption).[5] Land is the hidden resource that sustains every aspect of the economy, from food, through material products, furniture, buildings, to energy. Yet the EU is unaware of the real scale of our land consumption.

Previous studies revealed that in 2004, Europe's land footprint (including forestry) was already very high at 640 million hectares, 1.5 times the size of Europe itself[6] – and likely to increase further in line with consumption and plans to expand biofuels and bioenergy. Calculations of our land footprint from agriculture vary from 0.6[7] to 0.31[8] hectares per capita annually. By comparison, the UN recently

4. The principle of cascading use means biomass is used for materials and products first, and the energy content is recovered from end-of-life products, while respecting the waste hierarchy that requires reuse and recycling first.

5. FoEE (2013) Land footprint scenarios http://www.foeeurope.org/land-footprint-scenarios-041113; FoEE (2014) Meat Atlas http://www.foeeurope.org/meat-atlas

6. Sustainable Europe Research Institute (2011) Europe's global land demand http://www.foeeurope.org/sites/default/files/publications/Europe_Global_Land_Demand_Oct11%5B1%5D.pdf

7. Sustainable Europe Research Institute (2013) Hidden Impacts http://www.foeeurope.org/hidden-impacts-070313

8. United National International Resource Panel (2014) Assessing global land use http://www.unep.org/resourcepanel/Publications/AreasofAssessment/

Four footprints

The starting point is to measure and reduce our resource consumption, using the four footprints:

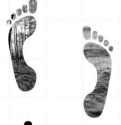

Land footprint: is the real area of land used, wherever it is in the world.

Water footprint: the total volume of water used, whether fresh water, rainwater or water polluted by the activity.

Carbon footprint: the total amount of climate changing gases released.

Material footprint: the tonnage of materials used, including the ore mined in order to extract metals.

suggested a target of 0.2 hectares of cropland per person per year, as an equitable global resource share.[9]

It is essential that the EU gets a clear picture of the amount of land it consumes, by measuring its land footprint, setting a reduction target and introducing policies that decrease this land dependency.

The four footprints: understanding the resource impacts of policies

The *Roadmap to a Resource Efficient Europe*, concluded that it is essential for the EU to measure land, water, material use and carbon emissions in order to better understand our resource consumption. Using four footprints – land footprint, water footprint, carbon footprint and material footprint (see illustration above) – provides a comprehensive overview of the overall resource use around the world, not just within Europe. Taking the water footprint of biofuels for example, it requires 2,500 litres of water to produce just one litre of biofuel.[10]

9. United Nations Environment Programme (2014) Sustainable consumption and production: targets and indicators. http://www.iisd.org/sites/default/files/publications/scp_targets_indicators.pdf

10. FAO (2009) Water information note http://www.fao.org/nr/water/docs/wateratfao.pdf

Friends of the Earth believes that the four footprints should be included in the 'EU Circular Economy Package', soon to be launched by the European Commission. This means measurement of Europe's resource consumption using the four footprints, and ambitious reduction targets for each of these. And it means using the four footprints in impact assessments for all EU policies. This will help decision makers to have consistency between policies, and to evaluate whether the introduction of a new policy would increase or reduce Europe's resource consumption.

By calculating the four footprints, clear links with the impacts caused by our consumption can be made. For example, some forms of bioenergy have a worse carbon footprint than the fossil fuels they replace. Increasing demand for biodiesel from palm oil, soy or rapeseed causes deforestation and habitat destruction, emitting significant quantities of greenhouse gases. Taking account of such 'indirect' greenhouse gas emissions gives a carbon footprint for crop-based biodiesel that can be worse than fossil diesel.[11] Similarly, scientists have shown that bioenergy from burning whole trees results in a carbon debt that is

11. IEEP (2011) http://www.foeeurope.org/Europe-biofuels-driving-destruction-101110

What should be done with this piece of land?

Grow biofuels.

Plant food crops.

Build housing.

Oops, sorry I'm late!

Fracking!

only repaid over decades, resulting in a net-climate impact that can be worse than coal combustion.[12]

Conclusions

Land is a vital but limited resource, one of Earth's nine planetary boundaries.[13] Europe's current demand for land to maintain our consumption-heavy lifestyles is not sustainable.

Bioenergy – from crops and primary wood in particular – has a substantial land footprint. Current EU plans to increase bioenergy production to generate a strategic share of EU energy imply a significant increase in Europe's global land consumption, and lead to competition with other land uses (including food and natural habitats) and with other regions.

Looking at the land footprint of Europe's bioenergy demand driven by EU policies and subsidies demonstrates the importance of measuring and setting limits to our resource consumption.

Current EU bioenergy policies do not take into account the land consumption required and the knock-on impacts, from land grabbing to biodiversity loss. Friends

12. RSPO & FOE (2013) Dirtier than coal http://www. rspb.org.uk/Images/biomass_report_tcm9-326672.pdf

13. http://www.nature.com/news/specials/ planetaryboundaries/index.html

of the Earth contends that 2008 EU targets for biofuels represent a policy failure which could have been avoided by conducting a full impact assessment including social impacts and a comprehensive assessment of resource consumption (based on the four footprints of land, water, carbon and materials).

It is vital in the next EU climate and energy package towards 2030 to bring together the renewable energy and resource efficiency agendas – to assess its overall resource consumption, and to ensure bioenergy is generated from sustainable sources.

Friends of the Earth calls on the EU to:

⇨ Set three ambitious and binding climate change, renewable energy and energy efficiency targets for 2030: greenhouse gas emissions must be reduced by at least 60% by 2030 with additional finance contributions to developing countries for climate action, and there must be binding targets to reduce energy use by 50% and increase the EU share of energy from renewables to at least 45%;

⇨ Introduce a cap to limit the use of bioenergy to sustainable levels and phase out crop-based biofuels and of the burning of whole trees for bioenergy;

⇨ Ensure efficient and optimal use of biomass resources for bioenergy, in line with the principle of cascading use; introduce comprehensive sustainability criteria covering environmental and social impacts so that only sustainable bioenergy is promoted;

⇨ Implement a comprehensive and mandatory bioenergy carbon accounting system that accounts for indirect land use change, carbon debt and indirect emissions from product substitution;

⇨ Refocus support towards bioenergy from wastes and agricultural and forestry residues, where indirect substitution emissions can be shown to be minimal. This would guarantee emissions reductions;

⇨ Make EU Member States report their land footprint annually – using a standard methodology and data – to be published alongside the material, water and carbon footprints, as per the *Resource Efficiency Roadmap*;

⇨ Put reduction targets in place in 2014 to help ensure the EU actively pursues the right policies to reduce its land footprint;

⇨ Introduce the measurement of land, water, materials and carbon footprints to EU and Member State impact assessments to enable the creation of policies that would reduce our resource consumption.

May 2014

⇨ The above information has been reprinted with kind permission from Friends of the Earth Europe. Please visit www.foeeurope.org for further information.

© *Friends of the Earth Europe 2014*

Don't write off biofuels yet, we will need them to get about in the future

An article from The Conversation.

By Adam Lee, Professor of Sustainable Chemistry at Aston University, and Karen Wilson, Professor of Catalysis and Research Director at Aston University.

Bioenergy and biofuels have an important role to play in lowering the use of carbon-intensive fossil fuels – a point underscored by the IPCC report which confirmed the need for further research to improve such technology.

A key challenge is creating alternative transport fuels, which are currently overwhelmingly fossil-fuel dependent, and responsible for 25% of greenhouse gas emissions in the EU. Of the renewable energy alternatives such as wind, tidal and solar power, it is only non-edible biomass (broadly, any biological matter derived from plants or other organisms) that can offer a low-cost, 'drop-in' sustainable transport fuel.

Such a liquid biofuel with a high energy density would fit easily into the enormous global fuel distribution networks that already exist. Other renewables such as wind and solar, though well-suited to powering static homes and industry, would require significant breakthroughs in battery technology before they could compete with gasoline, diesel or liquid biofuels.

With transport and emissions growing, an alternative is needed, fast. Biofuels derived from waste could replace 16%, or 37 million tonnes, of oil used by road vehicles in the EU alone by 2030. But history has shown how introducing new and transformative technologies can be slow, especially when they seek to usurp the highly-embedded infrastructure of the status quo.

New raw materials

Each year India produces more than 200 million tonnes of inedible agricultural waste such as rice and cotton stalks, unsuitable for either human consumption, animal fodder or bedding. Most is burned illegally to speed up the process of crop rotation. This releases huge quantities carbon dioxide into the atmosphere, and wastes valuable hydrocarbons that could be put to use. Europe produces 900 million tonnes of agricultural, forestry and food waste – all are rich in energy-filled sugars.

Which waste stream is used as a raw material feedstock for biofuels will differ from region to region, so a key challenge is to develop the technology that can process biomass with very diverse physical and chemical properties. For example, pine bark, switchgrass, corn husks and starches. Putting this to use would bring substantial benefits to industry and the economy: instead of paying to burn and bury biomass waste, companies could sell it as the starting point of creating valuable gasoline, diesel and jet fuels.

At the European Bioenergy Research Institute (EBRI) we are working on the engineering necessary. Solutions include thermochemical conversion, for distributed power generation such as to small farms and telecommunications towers. Thermochemical processes, such as pyrolysis, use heat (rather than burning) to force a chemical reaction. This can be dramatically accelerated and steered by the use of catalysts to lower the barriers to chemical bond breaking.

At high temperatures, the chemical bonds between carbon, hydrogen and oxygen atoms in biomass break. The products include a solid, called biochar (like charcoal, but not derived from coal), which is used as a soil enhancer or a solid heating fuel. It also produces a liquid that can be used as a biodiesel to power combustion engines, and a small amount of biogas that can be burned to sustain the pyrolysis heat reaction, introducing a degree of self-sufficiency that reduces the overall cost of the process.

Similar approaches pursued by private and state-funded businesses worldwide could deliver a double whammy of eliminating the carbon dioxide emissions from burning waste and at the same time yielding high energy density fuels. British Airways has this month committed to buy 50,000 tonnes of aircraft jet fuel derived from biowaste to be manufactured at a plant in Thurrock, Essex, as part of its deal with sustainable fuel company Solena. And one of the biggest champions of biofuels is the US military, currently offering US$4 billion in loans to companies able to help commercialise and push forward the technology required.

Modernising biofuels

It's important to end society's dependence on the use of non-renewable, carbon-based fuels. But this cannot be done at the expense of food crops (using corn to create ethanol, for example) or land used for food crops, as was the case with so-called first generation biofuels. The pressure this has put on food security in some parts of the world has blackened biofuels' reputation.

But it's crucial not to throw the baby out with the bathwater: the IPCC report urges us to explore the potential of all alternative energy sources. Newer, second generation biofuels can use non-edible plants (such as grasses) that can grow on soils that cannot sustain food crops. Better yet, careful selection and processing of waste biomass can accelerate the transition away from fossil fuels – we cannot simply write off biofuels as a credible solution to the world's future energy needs, as it has been by some. Bioenergy solutions are more practical, economic and immediate than many alternatives.

The authors of the IPCC report are absolutely right to highlight the urgent need to alter policy and investment strategies. Funding agencies, private industry and NGOs must work together to focus research and development efforts in bioenergy, and identify projects that will deliver rapid change, without compromising environmental and farming needs. The clock is ticking.

25 April 2014

⇨ The above information has been reprinted with kind permission from The Conversation Trust (UK). Please visit www.theconversation.com for further information.

Heather could be UK energy source

Heather could help meet bioenergy target and cut greenhouse gas emissions.

Harvesting energy from heather could cut greenhouse gas emissions and help the UK meet its bioenergy targets, according to new research by experts at Durham and Manchester Universities.

Heather grows naturally throughout the UK's uplands, but land managers often burn it back to create better feeding grounds for grouse and livestock.

The study, published in the journal *Biomass and Bioenergy*, says this releases energy equivalent to burning 36,000 tonnes of coal every year.

If all the UK's heather was harvested as a bioenergy crop, it could produce as much energy as 1.7 million tonnes of coal a year, achieving 15 per cent of the UK's 2020 biomass target without taking any farmland out of production.

'We have a large source of very efficient, low-carbon energy growing naturally on our uplands, but we're releasing all its CO_2 into the atmosphere without getting any energy from it,' says Professor Fred Worrall, from Durham University, who led the study.

'At the same time we're burning coal to satisfy demand, and that's a very dirty source of energy. Surely it would be possible to send the heather down the hill and stop burning the coal,' he adds.

The UK is committed to the EU policy of producing 20 per cent of its energy from renewables by 2020. A third of this is expected to come from biomass.

Elephant grass and short-rotation coppice are by far the most common bioenergy crops in the UK, but both require productive farmland which could otherwise be used to grow food.

By contrast, heather grows on largely unproductive heaths and moors. To assess its potential as a bioenergy crop, scientists burned samples in the lab and measured the energy released.

Using Countryside Survey estimates of heather coverage in the UK, they scaled their calculations up to the regional level, and subtracted the likely energy costs of harvesting and transporting the crop to incinerators.

As well as meeting a substantial part of the UK's bioenergy target, if it replaced its weight in coal, every hectare of heather could save 11 tonnes of carbon emissions a year.

But Worrall cautions that the industry could not take off everywhere. In parts of northwest Scotland, for example, heather grows too slowly, and the incinerators are too far away, to make it worthwhile.

Left to grow unabated, mature heather forms a blanket over the land. This makes poor grazing ground for sheep and grouse, and creates a large biological fuel store which can exacerbate the threat of wildfire.

For these reasons, land managers throughout the UK routinely burn off patches of heather. But the practice has attracted controversy in recent years.

There are concerns that it could be degrading moorlands, causing peat to erode more quickly into surrounding streams. This can harm water quality and raise the costs of treating it for human consumption.

If heather were to be harvested for energy, it would first need to be cut and baled. The environmental effects of this are not yet clear.

Worrall believes the main barriers to using heather as a bioenergy crop in the UK are now cultural, rather than technical.

'From an engineering perspective it's entirely possible,' he says. 'I think we have shown that there would be real benefits too.'

'But there is a culture for burning among land managers, and my suspicion is that this would be difficult to overcome.'

29 June 2014

⇨ The above information has been reprinted with kind permission from The University of Manchester. Please visit www.manchester.ac.uk for further information.

US Navy's solar panels in space could power entire cities (or wars)

Solar panel satellites, built in space by robots that beam power down to Earth – sound like science-fiction?

Well, even the team behind the idea admit it sounds 'nuts' but that's not going to stop them trying.

US Navy scientists are developing the project which in theory could power entire cities – or military endeavours.

The solar panels will be made up of two types of 'sandwich' module to form a one-kilometre-wide satellite.

Each module consists of a photovoltaic panel on top to absorb the Sun's energy, an electronics system in the middle to convert it to a radio frequency and a bottom antenna layer to beam the power back to Earth.

Dr. Paul Jaffe, a spacecraft engineer at the U.S. Naval Research Laboratory (NRL), said: 'It's hard to tell if it's nuts until you've actually tried.

'People might not associate radio waves with carrying energy, because they think of them for communications, like radio, TV or cell phones.

'They don't think about them as carrying usable amounts of power.'

The implications of successfully developing the technology are profound. Obviously it could solve many of our energy needs in an efficient and green manner.

But it could also enable a giant lumbering war machine – like the US Navy – to conduct global operations without the constraint of transporting and refuelling traditional fuels.

When you consider the Pentagon is the world's largest consumer of energy (excluding countries) this will be of particular interest to US military planners.

The technology is promising and has even spawned new ways of testing materials for space conditions.

Jaffe said: 'One of our key, unprecedented contributions has been testing under space-like conditions.'

Using a specialised vacuum chamber at another facility would have been too expensive, so Jaffe built one himself.

He said: 'It's cobbled together from borrowed pieces.'

The vacuum chamber is just big enough for one module. In it, Jaffe can expose the module to the simulated extreme cold of space and concentrated solar intensities (mimicked by turning on two powerful xenon lamps in the same spectrum as the sun).

By hooking the module up to a tangle of red and blue wires, he measures how well it radiates heat.

Jaffe says most solar panels orbiting with today's satellites are never tested in space-like conditions because the technology is already mature: 'But if you wanted to test anything under concentrated sunlight you would need something like the simulator we've put together here.'

Through trial and error, Jaffe has learned a lot. 'The capability we've built up with the testing and vacuum under Sun concentration is something that's pretty unusual.

'And we've actually gotten a couple inquiries from people who may want to use this.'

18 March 2014

Wave and tidal energy: part of the UK's energy mix

An explanation of the energy-producing potential of wave and tidal stream energy in the UK.

Overview

Wave and tidal stream energy is electricity generated from the movement of wave and tidal flows.

Wave power is much more predictable than wind power – and it increases during the winter, when electricity demand is at its highest. Tidal stream energy is also predictable and consistent.

It is estimated the UK has around 50% of Europe's tidal energy resource, and a study in 2004 estimated the UK's technical resource at around 16 terawatts per hour per year (TWh/year) (4% of overall supply).

Wave and tidal stream potential

Wave and tidal stream energy has the potential to meet up to 20% of the UK's current electricity demand, representing a 30-to-50 gigawatt (GW) installed capacity.

Between 200 and 300 megawatts (MWs) of generation capacity may be able to be deployed by 2020, and at the higher end of the range, up to 27 GWs by 2050 (see the *Renewable Energy Roadmap*).

The UK is currently seen as a world leader and focal point for the development of wave and tidal stream technologies because it has an abundance of marine energy resource.

With its excellent marine resource and its expertise in oil and gas exploration, the UK is in a unique position to benefit from this type of renewable energy – and to develop related wave and tidal stream services. The industry is still in its early stages however, and further research is needed to determine how best to exploit these assets.

Tidal range potential

Studies have estimated the UK's total theoretical tidal range resource at between 25 and 30 GWs – enough to supply around 12% of current UK electricity demand. The majority of this is in the Severn estuary (which has between 8 and 12 GW), with the estuaries and bays of the north west representing a similar amount and the east coast a further 5 to 6 GW.

The two-year cross-government Severn tidal power feasibility study could not see a strategic case for public investment in a Severn tidal scheme in the immediate term, though private sector groups are continuing to investigate the potential. Other potential projects assessed by developers at sites around the UK include the Mersey, the Solway Firth and the North Wales coast.

22 January 2013

⇨ The above information has been reprinted with kind permission from the Department of Energy & Climate Change. Please visit www.gov.uk for further information.

Hydro power scheme offers farming potential

Hydroelectricity can be an effective source of power without large reservoirs and damns and should be considered more widely on lowland as well as upland farms, says Oliver Routledge.

By Suzie Horne

'It is possible to generate significant amounts of income from relatively unobtrusive schemes that have little impact on the environment,' says Mr Routledge, who recently installed a 190 kW project on his family's 1,400-acre Scottish sheep and fish farm near Moffat in Dumfriesshire.

The scheme will earn in excess of £100,000 from the Feed-in Tariff (FiT) each year for the next two decades.

Some of the power generated will be used on the fish farm, saving it about £10,000 a year on its electricity bills, a saving that is expected to become more significant as conventional electricity costs rise.

'With fish food now in excess of £1,000/t, creating a stable diversified income stream to support a business that is so beholden to global commodity prices is very exciting and a great step forward for us,' says Mr Routledge.

In addition to the FiTs income and electricity cost savings, power is exported to the National Grid and earns an export tariff which is set in legislation and paid by the buyer of the power.

'For smaller schemes this is usually 4.64p/kW, but I have negotiated an initial price of 5.4p/kW with a green energy broker, worth around £45,000 a year,' says Mr Routledge, who also works for land agent Knight Frank's renewables and energy department.

The £750,000 cost of the hydro electric scheme, which includes the generating equipment and installation of 1,800m of pipes from a burn 170m above the steading, should be paid off within five to six years.

Three years passed between the point at which the decision was made to invest in the scheme to it being installed and working. Two years would usually be a reasonable timescale for an installation of this type, says Mr Routledge, to allow for a year's water flow monitoring, planning, grid connection, construction and commissioning. It was grid connection complications which held things up in this case.

The amount of untapped hydropower in Scotland is estimated to be the equivalent of 20,000 similar schemes.

FiT payments for renewable technologies are reviewed annually and those for new hydropower schemes are likely to be cut by 2.5 to 5% from April next year.

'Even in lowland areas, there is the potential to install many more small hydro projects. These do not require large differences in height to create a head of water, but rely instead on a higher volume of slower-moving water to power the likes of an Archimedes screw,' says Mr Routledge.

3 December 2013

⇨ The above information has been reprinted with kind permission from *Farmers Weekly*. Please visit www.fwi.co.uk for further information.

Eric Pickles' war on windfarms could blow away bid to hit emissions targets

An article from The Conversation.

By David Elliott, Emeritus Professor of Technology Policy at The Open University

THE CONVERSATION

Do we have enough onshore windfarms, or do we have too many? And who decides what 'too many' looks like? The Conservative Party has announced it would end subsidies for new onshore wind farms if it won the next election, and Communities Secretary Eric Pickles has demonstrated an aggressive approach that has seen two thirds of planning applications for onshore wind farms rejected in the last year – 164 projects halted since January.

For example, under the ministerial power to call in a planning decision, Pickles rejected the plan proposed jointly by Wind Prospect and EDF for a six turbine, 12 MW wind farm near Bridlington, Yorkshire – one of ten onshore wind farms rejected by call-in over the past year – in doing so overturning the decision of a planning inspector.

Is this war on windfarms wise? Tory peer Lord Deben was reported as saying the UK had 'enough' onshore wind projects, and that though they will continue to be important after 2020, 'it is not a decision we need to make now'. The Committee on Climate Change which he chairs reiterated that increasing onshore wind capacity during the 2020s was likely to be the most cost-effective means for the UK to meet its carbon emissions targets.

According to the Committee, while there is sufficient capacity online and in the pipeline to meet 2020 targets, this relies on there being sufficient investment in other renewable technologies. Beyond 2020, there will be choices to be made on how to decarbonise Britain's energy generation system. The most cost-effective method entails more onshore wind development through the 2020s; doing anything else faces raising costs with knock-on effects on energy prices and affordability.

Clearly the actual outcome will depend a lot on politics. Trade association Renewable UK stated, rather bitterly, that it is 'unfortunate that we've reached a point where the Conservatives are allowing UKIP to dictate Tory energy policy', calling Tory efforts to mollify the party's rural electoral heartlands ahead of next year's election 'ill-considered, short-term policy-making on the hoof' and likely to damage what is becoming a flagship British industry.

According to the Committee on Climate Change, the most cost-effective means to hit the emissions reductions targets for 2020 is 15 GW of onshore wind up and running by then, rising to 25 GW by 2030 – on the basis that it is 'relatively low cost'. But if not that, then what? The UK already has around 10 GW of wind capacity on and offshore, which according to National Grid needs to more than double to 26 GW by 2020. If onshore wind projects, the far easier and cheaper of the two, are to be slowed or even halted, the task becomes considerably more difficult.

It will be made even harder still if the Government persists with its plan to halt funding for large solar farms. That's another technology Pickles has targeted, although his rejection of one in Suffolk has been overturned by judicial review.

Given policies like this, David Toke at Aberdeen University suggests that the UK will fall far short of its target for 15% energy generated from renewables by 2020 agreed with the EU.

This has all come about despite public support for onshore wind farms in the UK remaining high. In the last survey conducted by the Department for Energy and Climate Change, 70% were in favour. Public support for solar was even higher – 85%. Yet for specific projects, objections are raised about intrusion and, more prosaically, in the case of wind farms, their possible impact on house prices.

Another issue for wind is noise, and in particular infrasound. While audible noise even at very low levels can be annoying, which means the exact location of a wind farm is key, the claim that significant health problems arise from ultrasound seems to have been roundly disproved.

Among critics, much is made of the problems stemming from over-reliance on wind and solar, which by their nature are intermittent, depending on the weather. However, in the case of wind, a Royal Academy of Engineering study concluded there would be no major problems integrating intermittent energy from wind into the grid until it contributed more than 20% of the national demand. Once major contributions from offshore wind and other renewables, including solar, pushed their contribution beyond 20%, new grid balancing strategies will be needed, but there are plenty of options.

It's not technology that's the problem. While there are disagreements on the role of nuclear power, the relative merits of renewables and the subsidies they both require, all agree that onshore wind is the cheapest non-fossil fuel energy source. The question is whether nimbyish responses to the changing face of energy generation are allowed to overshadow what agreement there is.

25 June 2014

Wind farm expansion will require fridges to be switched off at times of peak demand

Britain's increasing reliance on wind power in the 2020s could cause power shortages without a radical overhaul of the grid, including automatic control of household appliances.

By Emily Gosden

Household fridges and freezers will need to be automatically switched off at times when Britain's electricity demand is high, in order to keep the lights on as Britain becomes more reliant on wind energy, experts say.

The current electricity grid will struggle to cope with the number of wind farms expected to be built by the early 2020s because the power they produce is so intermittent, according to a report from the Royal Academy of Engineering.

A radical overhaul of the way the electricity system is managed – including a 'smart grid' that can control household appliances to reduce demand when power supply is inadequate – will be needed, it finds.

Britain will also need to build more power import and export cables to the continent to help manage variable wind power output, and develop storage technologies to keep surplus power for times when there is a shortfall.

The measures will be necessary to avert blackouts under a vast expansion of wind power – unless Britain instead builds an expensive

new fleet of reliable power stations to be fired up as back-up when the wind doesn't blow, it found.

Professor Roger Kemp of Lancaster University, said: 'If there is a sudden peak in electricity demand the smart freezer would say, "OK I'm going to switch off for half an hour until that peak is over".'

Consumers could negotiate cheaper energy tariffs for consenting to this and would not be affected by the temporary switch-off, he said.

'If you didn't have a smart grid and smart control of domestic equipment, you would probably find prices would have to go up more because the power-system people would have to build more power stations as back-up.'

The report says that 'the ability to manage demand to reflect the output from wind will be vital to the successful integration of larger amounts of wind capacity'.

However, it casts doubt on the viability of this solution, warning: 'There is much uncertainty on how effective it will be and at what cost.'

The report finds that the current grid can cope with 'up to a 20pc

contribution from wind power without the need for significant upgrades to the system and using existing balancing mechanisms'.

Beyond that threshold – expected to be crossed by the early 2020s – 'managing the system will become increasingly difficult', it warns.

The inherently variable nature of wind power will present problems when there is a mismatch between wind output and consumer demand – either too much wind or not enough.

'During periods of high demand, wind still often produces very low levels of output,' the report finds.

'At low levels of penetration [wind energy deployment], this should not be a major issue and, indeed, up to now security has not been compromised despite periods of virtually no output from wind and maximum demands. However, as levels of penetration increase, the situation can be expected to change adversely.'

If wind power continues to expand beyond 2020, days of negligible wind power could 'present problems for security of supply'.

In the reverse scenario, where wind output is high but demand is low, wind farm owners could increasingly have to be paid to switch their turbines off – an 'inefficient and costly' solution if it becomes more common.

10 April 2014

⇨ The above information has been reprinted with kind permission from *The Telegraph*. Please visit www.telegraph.co.uk for further information.

Could burning coal under the sea provide 200 years of 'clean' energy?

By Mat Hope

The Government claims it's on a mission to clean up the UK's energy system. You could be forgiven for thinking that means an end to coal power – the most polluting energy source of all. But, thanks to new technology, the Government hopes there's a new, 'clean' way to keep using coal.

Writing in *The Telegraph* this weekend, Algy Cluff, chief executive of energy company Cluff Natural Resources, says 'underground coal gasification' could 'provide a vital energy solution and produce abundant and cheap gas for generations'. The prospect has piqued the Government's interest, and energy minister Michael Fallon has established a working group to explore its feasibility.

But is it too good to be true?

What is underground coal gasification?

Underground coal gasification (UCG) involves drilling down into coal – normally deep underground – then igniting it. The resulting gas then runs up another borehole and is collected on the surface.

Once the gas is collected, companies can use it to run power stations, or convert it into transport fuel. Carbon capture and storage (CCS) technology can be added, reducing the process' emissions, and making it relatively 'clean'.

As such, the Government now sees the 'exciting potential' of UCG as means to generate abundant, domestically-sourced, ostensibly fairly low carbon power in the UK, *The Telegraph* reports.

So, what are UCG's prospects in the UK?

200 years of coal

UCG is normally conducted onshore, but new 'horizontal drilling'

technology means it could now be possible to get at hard-to-reach offshore coal seams. But could it really provide energy for 200 years, as *The Telegraph* claims?

The Telegraph cites British Geological Survey (BGS) research which suggests there could be as much as 17 billion tonnes of coal located off Britain's shores. It says that could be enough to provide energy for 289 years, based on 2004 consumption levels.

But that's assuming all the resources can be accessed and are fit to undergo UCG. BGS analyst, Nigel Smith, tells us there are still a lot of unknowns.

He says that while horizontal drilling technology does make it easier to access offshore coal resources, it's still not clear where suitable coal seams might be located. He acknowledges that some areas which BGS identified as potential UCG locations 'could be quite controversial'.

Even in areas where there is existing data, Smith says there's 'quite a bit of work to do' to gather data, with some places still being a total blindspot. Companies will need to drill more boreholes to get a better idea of where suitable coal seams are, he says – which takes time and investment.

There could also be public opposition to UCG, just as there has been to shale gas in recent months, further slowing progress. Though Julie Lauder from industry group, the Underground Coal Gasification Association, says because the UK is predominantly looking at using UCG offshore, this could be less of an obstacle to the industry.

The technology is already dividing opinion, despite only being in its infancy, however. Lauder says the process is akin to scaling up the coal fires some people have in their front rooms. Campaign group, Frack Off, sees it somewhat differently, however. It warns the process would create what it describes as 'hell on Earth', as it involves sparking large fires deep underground.

Once the technical and social obstacles are considered, it's very hard to estimate exactly how much coal could end up being gasified.

Emissions

UCG could be a low-carbon form of power generation if CCS technology – which is currently unproven on a large scale – was added. That in turn affects the technical and economic viability of the projects.

As the Climate Change Act legally obliges the Government to reduce the UK's emissions by 80 per cent by 2050, the onus will be on the UCG industry to prove the low emissions technology works. If it can't show UCG with CCS is viable, it's questionable whether or not the projects would be allowed to proceed.

Michael Blinderman, director of Ergo Exergy Technologies, which develops UCG equipment, estimates that UCG power plants with CCS could emit 26 per cent less carbon dioxide than some natural gas generators. His calculations suggest UCG gas could emit as little as 380 grams of carbon dioxide per kilowatt hour of electricity generated.

That's pretty good going compared to other fossil fuels, though it's still more than nuclear and renewables.

But CCS is yet to be proven on a large scale.

The Government has just given £1 billion of funding to kick start research programmes, and the Environment Agency requires any UCG power plant to have CCS fitted, but it's unclear when such projects could become a reality. And if the UCG gas gets turned into transport fuel, companies don't have to fit any CCS technology – making it much more carbon intensive.

Questions also remain over where the carbon could be stored, if it's captured. About 30 per cent of the captured carbon dioxide could be pumped back into the coal chamber, Lauder tells us – but that still leaves 60 per cent to store.

So while UCG could be made into a low-carbon form of power generation with the addition of CCS technology, it's unclear when the technology will exist to make that a realistic prospect.

16 December 2013

⇨ The above information has been reprinted with kind permission from The Carbon Brief. Please visit www.carbonbrief.org for further information.

© The Carbon Brief 2014

'Poo power': UK firm turns human waste into clean energy

A UK power company has revolutionised the supply of gas to homes in a method that extracts energy directly from human waste.

Severn Trent Water (SVT) claims to be the first power company in the UK to provide biomethane gas, created by breaking down the 'sludge' in Britain's sewers. The sludge is a combination of human waste compounds and biodegradable matter.

The cocktail of waste is then cleaned and compressed, before tests ensure it performs like regular gas. Human waste can also be broken down into carbon dioxide, which can then be further refined into biomethane.

SVT, Britain's second largest publicly traded water company, began sending the natural gas to the National Grid last week, marking the first time any kind of eco-friendly gas has been delivered for public use. The biomethane is developed in SVT's Minworth treatment plant, a facility which cost £8.4 million (US$13.6 million) to build and can develop clean gas for over two million people.

'Local domestic customers will be able to tap into the energy contained within the biogas and biomethane as it's injected into the grid,' said SVT's renewable energy development manager Simon Farris.

While Farris admitted to the BBC that the source material was 'a little unsavoury', he said the new method could cut SVT's carbon emissions by 300,000 tons over the next 20 years.

Other energy producers including Wessex Water and Northumbrian Water also have plans to develop a pipe to supply natural gas from sewers.

'Greenhouse-gas emissions reductions could be significant, as the methane normally generated at sewage works is 25 times more harmful to the atmosphere than carbon dioxide,' Dragan Savic, a professor of hydroinformatics, told *The Independent*.

'By capturing methane and pumping it into the National Grid, water companies could turn from greenhouse-gas emitters into renewable-energy generators,' Savic said.

There are over 9,000 sewage treatment plants in the UK, although none are used to provide energy to the public.

According to data from the Environment Agency, more than four million tons of greenhouse gases are emitted by the water treatment industry alone, including high concentrations of nitrous oxide – a gas claimed to be almost 300 times more dangerous than carbon dioxide.

In the past, energy companies have used the method to provide electricity and heat on site, but advances in cleaning technology now mean high-quality biomethane gas may be the way forward for UK households in the near future.

2 October 2014

⇨ The above information has been reprinted with kind permission from RT. Please visit www.rt.com for further information.

© Autonomous Nonprofit Organization 'TV-Novosti', 2005-2014

Key facts

- A commitment to renewable energy is on the world political agenda, with Europe, for example, proposing a target of increasing total use of renewable energy from 7% to 20% by 2020. (page 1)

- Coal has the most widely distributed reserves in the world and is mined in over 100 countries. (page 1)

- In an average year, a typical coal power station generates 3,700,000 tons of carbon dioxide, and is the primary human cause of global warming – that's as much carbon dioxide as cutting down 161 million trees. (page 1)

- Oil is the most commonly reported cause of water pollution, with over 5,000 incidents recorded annually by the Environment Agency. (page 1)

- The UK possess 40% of Europe's total wind resource. (page 1)

- The photovoltaic cells used in the process of solar harnessing only currently absorb around 15% of the sunlight's energy. (page 2)

- Electricity generated from renewable sources in the UK in 2013 increased by 30 per cent on a year earlier, and accounted for 14.9 per cent of total UK electricity generation, up from 11.3 per cent in 2012. (page 4)

- UK households spent an average of £106 a month on household energy in 2012. This was a 55% rise on the 2002 monthly spend, after accounting for inflation. This is despite a decline in average energy usage. (page 7)

- On average, households spent the equivalent of 5.1% of their income on household energy in 2012, up from 3.3% in 2002. Most of this rise occurred between 2004 and 2009. (page 8)

- Almost seven in ten (68%) say they have turned the heating down or off when they ordinarily would have left it on, 27% have spent less on food while 5% have borrowed from short term lenders in order to fulfil bills. (page 9)

- In terms of changes to the market, 69% favour forcing energy supplies to reduce the number of different tariffs they offer and simplify bills. (page 9)

- A survey commissioned by the Energy Saving Trust revealed that 48 per cent agreed that it's a hassle to change energy suppliers. (page 10)

- A survey commissioned by the Energy Saving Trust revealed that two thirds (66 per cent) of people think that more heat is lost through the roof of their home than the walls – however, for the majority of properties the walls will actually lose more heat. (page 10)

- Energy Saving Trust's survey found that just 27 per cent incorrectly said turning up their thermostat to a high setting heats the home faster, compared to 54 per cent that correctly said the statement was false. (page 10)

- On average desktop computers cost around £24 a year to run. (page 11)

- The UK has 19% of households in energy poverty. (page 13)

- A report from Anglia Ruskin University reveals that by, using the most recent data on known reserves and current consumption, the UK has approximately 5.2 years of oil, 4.5 years of coal and three years of gas remaining. (page 18)

- Government figures suggest that, through to 2030, the net benefit of the smart meter programme to the UK will total £6.2 billion. (page 19)

- According to research carried out by the Smart Meter Central Delivery Body, 84% of people had heard of smart meters, however, just 44% said they would be interested in having one installed. (page 21)

- Fracking: a survey by Opinium for Govtoday has revealed that 37% of adults are opposed to the exploitation of shale gas reserves while only 25% are in favour. However, as many as 28% describe themselves as neither in favour nor opposed to the controversial drilling technique, and another 10% 'don't know'. (page 23)

- Nuclear energy has supplied up to a third of the UK's electricity since 1956. Today 18 reactors on nine sites supply nearly a fifth of the country's electricity. (page 24)

- In the UK, the power generated by existing power stations avoids the emissions of 40 million tonnes of carbon dioxide a year – the equivalent of taking almost half of Britain's cars off the road. (page 25)

- The land footprint of bioenergy consumption in 2010 for Europe was 45 million hectares, equivalent to the entire land area of Sweden. (page 28)

- Biofuels derived from waste could replace 16%, or 37 million tonnes, of oil used by road vehicles in the EU alone by 2030. (page 31)

- Wave and tidal stream energy has the potential to meet up to 20% of the UK's current electricity demand. (page 34)

- Two thirds of planning applications for onshore wind farms rejected in the last year (2013-14) – 164 projects halted since January. (page 36)

Biofuels and biomass

Plants use photosynthesis to store energy from the Sun in their leaves and stems. Living things, like these plant materials, are known as biomass. The wide range of fuels derived from biomass are known as biofuels. Corn ethanol, sugar ethanol and biodiesel are the primary biofuels markets.

Energy

A force which powers or drives something. It is usually generated by burning a fuel such as coal or oil, or by harnessing natural heat or movement (for example by using a wind turbine).

Fossil fuels

Fossil fuels are stores of energy formed from the remains of plants and animals that were alive millions of years ago. Coal, oil and gas are examples of fossil fuels. They are also known as non-renewable sources of energy, because they will eventually be used up: as they are finite, once they are gone we will be unable to produce more of them.

Fuel cell

A fuel cell is an electrochemical device which, at its simplest, converts hydrogen and oxygen into water, along the way generating electricity. As such, fuel cells can be used to power any device which requires electricity.

Fuel poverty

A household is said to be in fuel poverty if they spend more than 10% of their income on heating their home.

Geothermal power

The Earth is hot inside. Most of this heat comes from radioactive decay, which heats up the surrounding rocks. This heat can be used as an energy source: water is pumped down into the hot rocks, and the steam produced used to drive an electricity generator. The hot water can also be used directly to heat homes and businesses. However, unless the rock conditions are just right, geothermal power is not cost-effective.

Hydropower (or water power)

Energy which is generated using the movement of running water. This includes tidal/wave power.

Microgeneration

Microgeneration is the production of heat or electricity by individual households and small businesses. Microgeneration technologies are low- or even zero-carbon and allow householders and business owners to generate their own sustainable heat and/or electricity.

NIMBYism

NIMBY stands for 'Not In My Back Yard'. It refers to the attitudes of local residents who oppose developments such as nuclear power stations or wind farms in their area.

Nuclear power

A method of generating energy using controlled nuclear reactions. These are used to create steam, which then powers a generator. Nuclear power is controversial and subject to much debate, with proponents saying it is a greener and more sustainable alternative to fossil fuels, whereas opponents argue that nuclear waste is potentially hazardous to people and the environment.

Offshore wind farm

An offshore wind farm consists of a number of wind turbines, constructed in an area of water where wind speeds are high in order to maximise the amount of energy which can be generated from wind.

Renewable energy

Energy generated from natural resources such as wind, water or the Sun. Unlike fossil fuels, energy can be generated from these sources indefinitely as they will never run out.

Solar power

Energy generated by harnessing the heat of the Sun.

Wind power

Energy which is generated using movement powered by wind. This is most commonly achieved via wind turbines, which are used to produce electricity.

Winter Fuel Payment

This is a yearly tax-free payment provided by the Government to help older people pay for their heating in the winter. It is available to UK residents over 60 years of age.

Assignments

Brainstorming

⇨ In small groups, discuss what you know about energy and energy alternatives. Consider the following points:

- What is renewable energy?
- What is non-renewable energy?
- What does it mean to be 'energy efficient'?
- What is fuel poverty?
- What is a Winter Fuel Payment?
- What are the four different footprints and how do they relate to energy?

Research

⇨ Research possible sources of renewable energy that could be used instead of fossil fuels in the future. You can use the articles in this book as well as doing further research using the Internet. In your opinion, which is the most viable energy source? Write a short summary of your findings and conclusions based on the research you have carried out.

⇨ Devise and carry out a survey among your year group to assess their attitudes towards different types of energy. It should contain no more than ten questions and you should have at least 15 respondents. Do your results show much diversity in people's opinions?

⇨ How much electricity do you use in a day? Keep an energy diary over the course of one day, recording every time you switch on a light, boil a kettle, watch television etc. and how long before you turn each device off again. Are you surprised by how much you rely on electricity? Come up with one way of reducing your energy consumption during the day.

⇨ Choose a country from a map and research further their energy situation. How do they generate and produce energy? How much is generated from renewable and non-renewable sources? How much do they consume? How much does the country produce itself? How dependent are they on energy imports? Create a PowerPoint presentation that summaries your findings.

Design

⇨ Design a poster which helps bust energy saving myths in a bid to help UK householders cut fuel bills. You might want to include commonly incorrect statements and correct them and/or tips on how to be energy efficient. You might find reading *Half of UK householders think it's cheaper to leave the home heating on all day at low temperature* on pages 10–11 helpful.

⇨ Create a leaflet that will inform UK homes about the rollout of the smart meter programme. Include information about what a smart meter is and its potential advantages and disadvantages.

⇨ Choose one of the articles in this book and create an illustration to highlight the key themes/message of your chosen article.

Oral

⇨ Imagine a proposal has been put forward for a wind farm to be built in your local area. Role play a local council meeting, with one group of students acting as the councillors supporting the proposal, and the other group of students acting as some local residents who oppose the proposal. Could the councillors be accused of not taking residents' concerns into account? Could the residents be accused of 'NIMBY'ism? What might be their concerns?

⇨ As a class, discuss fracking. What is it? Why is it so controversial? Has it been utilised successfully anywhere? Or not so successfully? Read the articles *Fracking threat to the UK* on page 22 and *Poll shows 63% not opposed to fracking* on page 23 and discuss the authors' points of view.

⇨ Research all the different energy supplies in the UK and create a five-minute presentation. What kind of different tariffs do they offer (e.g. do they offer fixed price energy?)? How do they compare? You might find www.uswitch.com helpful.

⇨ As a class, discuss the following statement: 'Bioenergy and biofuels play a key role in the future of energy.'

Reading/writing

⇨ Read the article *UK future energy scenarios 2014: summary* on pages 16–17 and write a summary for your school newspaper.

⇨ Write an article exploring fuel poverty in the UK. Who does it mean? Who does it effect? How can we help solve this problem?

⇨ Watch *Avatar* (12a) and write a review exploring how the director deals with the theme of energy.

⇨ Read *The Windup Girl* by Paolo Bacigalupi and write a review exploring how the author deals with the theme of energy.

Acknowledgements

The publisher is grateful for permission to reproduce the material in this book. While every care has been taken to trace and acknowledge copyright, the publisher tenders its apology for any accidental infringement or where copyright has proved untraceable. The publisher would be pleased to come to a suitable arrangement in any such case with the rightful owner.

Images

All images courtesy of iStock, except page 15: MorgueFile, page 20: icons by Freepik, page 25: icons by Freepik, Stephen Hutchings, Yannik, Icomoon, OCHA from www.flaticon.com, page 29: forest © Joshua Mayer, water ripple © Griffin Keller, ore/mineral © James St.John, page 33: MorgueFile, pages 34 & 35 © Kundan Ramisetti, page 37 © John Hughes and page 38 © Jackie Staines.

Illustrations

Don Hatcher: pages 5 & 22. Simon Kneebone: pages 3 & 17. Angelo Madrid: pages 8 & 30.

Additional acknowledgements

Editorial on behalf of Independence Educational Publishers by Cara Acred.

With thanks to the Independence team: Mary Chapman, Sandra Dennis, Christina Hughes, Jackie Staines and Jan Sunderland.

Cara Acred

Cambridge

January 2015